I0465483

Voices for the Earth

INSIDE COP29

Shaping the Future of Climate Diplomacy

Sophia Reynolds

All rights reserved. No part of this publication may be reproduced, distributed, or transmitted in any form or by any means, including photocopying, recording, or other electronic or mechanical methods, without the prior written permission of the publisher, except in the case of brief quotations embodied in critical reviews and certain other noncommercial uses permitted by copyright law.

TABLE
OF CONTENTS

CLIMATE CHANGE

The **future of our planet** hinges on our collective ability to respond to the climate crisis. As the **effects of climate change** continue to intensify - through rising temperatures, extreme weather events, and widespread environmental degradation - the need for **global action** has never been more urgent. The **Conference of the Parties (COP)** has become the foremost arena for international climate diplomacy, where world leaders, negotiators, scientists, activists, and businesses gather to chart a course for a sustainable, low-carbon future. The **COP29** conference marks another critical chapter in this ongoing journey, one that will set the stage for how we confront the climate emergency in the coming decades.

This book aims to provide a comprehensive overview of **COP29**, the global negotiations that will shape the future of climate policy, and the key factors influencing climate action on the world stage. As we stand at the crossroads of history, the decisions made at COP29 have the potential to steer the world toward a **climate-resilient future** or further delay critical action. From **climate justice** to **renewable energy innovations**, **financial mechanisms** to **carbon markets**, COP29 will see a convergence of ideas and commitments that can either propel us forward or set us back in our fight to combat climate change.

In this book, we will delve into the intricacies of **climate diplomacy**, exploring the **global partnerships** formed, the **policy frameworks** that guide negotiations, and the **challenges** and **opportunities** that arise when nations with varying priorities and capacities come together to address a shared global threat. We will also examine how **emerging technologies**, **finance mechanisms**, and **international cooperation** can drive climate solutions. Alongside these discussions, we will focus on the **critical voices** of vulnerable nations, **youth movements**, and **local communities**, whose futures depend on the success of these global efforts.

The significance of **COP29** goes beyond the technicalities of international negotiations. It represents a **decisive moment** in our collective journey toward **climate action**, demanding not only political will but also a commitment to equity, justice, and long-term sustainability. As we explore the details of COP29, we must ask ourselves: will this be the moment we take meaningful action, or will we look back and see another missed opportunity?

This book is not just a chronicle of the negotiations that take place in **Dubai**, but a call to action. It is an invitation for everyone—policymakers, businesses, activists, and citizens—to engage with the climate crisis, understand its far-reaching impacts, and contribute to the global solution. The time to act is now. The stakes are high. The path we take in the coming years will define the world we leave behind for future generations.

Let us look ahead with determination, innovation, and the collective strength to ensure a thriving, sustainable world for all. The future of climate diplomacy starts with us, today.

GLOBAL
CLIMATE ACTION

Chapter 1

Introduction to COP29

Overview of COP (Conference of the Parties)

The Conference of the Parties (COP) is an annual international climate summit convened under the United Nations Framework Convention on Climate Change (UNFCCC). COP was established in 1992 at the Earth Summit in Rio de Janeiro, where the UNFCCC was initially adopted. The primary purpose of COP is to serve as the central platform for countries around the world to coordinate, negotiate, and implement global strategies addressing climate change. Each conference brings together representatives from nearly 200 countries, including government officials, scientists, environmental organizations, and other key stakeholders, making COP the most comprehensive climate governance event in the world.

Objectives of COP

The overarching goal of COP is to prevent dangerous human interference with the climate system. This objective is pursued through:

1. **International Negotiations**: The conference is a forum for countries to negotiate emissions reduction targets, financial commitments, and cooperative measures.

2. **Binding Agreements**: COP has led to legally binding agreements, like the Kyoto Protocol in 1997 and the Paris Agreement in 2015, which marked significant milestones in global climate policy.

3. **Climate Adaptation and Resilience**: In addition to mitigation, COP discussions increasingly focus on adaptation strategies for countries facing severe climate impacts, particularly vulnerable island states and developing nations.

4. **Climate Finance**: A major role of COP is mobilizing and regulating funding mechanisms to support climate initiatives worldwide, including the Green Climate Fund (GCF) established to aid developing nations.

Evolution of COP

Since the first COP in Berlin in 1995, each conference has built on previous negotiations, evolving in focus and ambition as the scientific understanding of climate change has expanded. Initially, COP focused on securing consensus around the necessity of limiting greenhouse gas emissions, which led to the establishment of differentiated responsibilities for developed and developing nations. As climate impacts became more apparent, COP agendas began to include not only mitigation but also adaptation, resilience, and financing. COP21 in Paris marked a turning point, achieving the historic Paris Agreement, where countries committed to limiting global temperature rise to well below 2°C above pre-industrial levels, with efforts to keep it to 1.5°C.

Key Agreements from Past COPs

1. **Kyoto Protocol (COP3, 1997)**: The first legally binding agreement that set emission reduction targets for developed countries, introducing concepts like carbon trading.

2. **Paris Agreement (COP21, 2015)**: A landmark global agreement to limit warming to 1.5°C–2°C, it introduced Nationally Determined Contributions (NDCs) for countries to set and review their climate commitments every five years.

3. **Glasgow Climate Pact (COP26, 2021)**: Emphasized the urgent need to close the emissions gap and mobilized greater financial resources, focusing on phase-down of coal and methane reduction.

Significance of COP in Global Climate Policy

The COP conferences serve several vital functions that contribute to the international climate policy landscape:

1. **Setting and Enforcing Global Climate Goals**: COP is unique in its ability to set ambitious targets for reducing greenhouse gas emissions and advancing climate resilience, creating a framework for international accountability.

2. **Fostering International Cooperation**: By bringing together nearly all countries, COP enables global coordination, encouraging collaboration on technology transfer, shared knowledge, and joint adaptation and mitigation efforts.

3. **Mobilizing Financial Resources**: COP plays a critical role in organizing financial support for developing countries and vulnerable regions to implement climate action, including renewable energy projects, disaster preparedness, and infrastructure resilience.

4. **Influencing National Policies**: The commitments made at COP have a direct impact on national climate policies, as countries incorporate their NDCs into domestic laws, incentivize green technology, and set carbon neutrality goals.

5. **Supporting Science-Based Climate Action**: COP reinforces the connection between climate science and policy, grounding decisions in the latest research from the IPCC and other scientific bodies. This has helped raise awareness of the urgency of climate action and the need for evidence-based policies.

Challenges Faced by COP

Despite its achievements, COP faces challenges in meeting its goals, including:

1. **Political Tensions**: Conflicting national interests, particularly between developed and developing countries, can stall progress, as seen in disputes over financing, emissions cuts, and responsibilities.

2. **Financing Gaps**: While COP has established financing mechanisms, there remain significant funding gaps, especially in fulfilling pledges to developing countries.

3. **Non-Binding Nature of Some Commitments**: Many of the agreements reached are non-binding, relying on voluntary compliance, which can lead to gaps in accountability.

Looking Forward: The Future Role of COP

The importance of COP continues to grow as the climate crisis intensifies. COP's role in strengthening global climate governance, increasing climate finance, and setting ambitious climate targets remains crucial. With each new conference, COP is expected to push for deeper commitments, innovative solutions, and inclusive decision-making that incorporates the voices of indigenous communities, youth, and other underrepresented groups. The future of COP will likely involve enhancing adaptation measures, securing long-term financial support, and fostering resilience in the face of mounting climate challenges.

Historical Milestones Leading Up to COP29

Here are some key historical milestones that led to COP29, marking significant developments in international climate policy and shaping the framework within which COP29 operates:

1. 1972: United Nations Conference on the Human Environment (Stockholm Conference)

- This was the first major international conference addressing environmental issues on a global scale. It led to the formation of the United Nations Environment Programme (UNEP) and laid the groundwork for subsequent climate and environmental initiatives.

2. 1988: Establishment of the Intergovernmental Panel on Climate Change (IPCC)

- The IPCC was created by UNEP and the World Meteorological Organization (WMO) to provide scientific assessments on climate change, its impacts, and potential response strategies. The IPCC's reports have been pivotal in informing global climate policy and negotiations at COPs.

3. 1992: United Nations Framework Convention on Climate Change (UNFCCC)

- Adopted at the Earth Summit in Rio de Janeiro, the UNFCCC established a framework for international cooperation to combat climate change by stabilizing greenhouse gas concentrations. It recognized the need for differentiated responsibilities between developed and developing countries, setting the stage for future COP conferences.

4. 1997: Kyoto Protocol (COP3, Kyoto, Japan)

- COP3 marked the adoption of the Kyoto Protocol, the first legally binding treaty on greenhouse gas emissions. The protocol committed developed countries to specific emissions reduction targets, introducing mechanisms like emissions trading, the Clean Development Mechanism

(CDM), and Joint Implementation (JI) to promote cost-effective climate action.

5. 2009: Copenhagen Accord (COP15, Copenhagen, Denmark)

- COP15 sought to establish a successor to the Kyoto Protocol, but the conference was marred by disagreements between developed and developing nations. The Copenhagen Accord, a non-binding agreement, introduced the idea of limiting global temperature rise to 2°C and initiated significant climate financing pledges, aiming for $100 billion annually by 2020 for developing countries.

6. 2015: Paris Agreement (COP21, Paris, France)

- COP21 achieved a landmark moment with the adoption of the Paris Agreement, a legally binding accord that committed nearly all countries to work toward limiting global warming to well below 2°C, with a target of 1.5°C. It introduced Nationally Determined Contributions (NDCs), requiring countries to submit their climate targets and strengthen them every five years. The Paris Agreement underscored global solidarity and signaled a shift toward long-term, collective climate action.

7. 2018: Katowice Climate Package (COP24, Katowice, Poland)

- COP24 focused on creating the "rulebook" for the Paris Agreement, clarifying how countries should report and track their emissions, climate efforts, and NDCs. The Katowice Climate Package laid the groundwork for transparency, accountability, and the monitoring of global progress.

8. 2021: Glasgow Climate Pact (COP26, Glasgow, Scotland)

- COP26 made notable progress by committing to "phase down" coal, a first for COP language, and underscoring the urgency of limiting warming to 1.5°C. The conference emphasized reducing methane emissions, increasing climate finance, and establishing dialogue on "loss and damage" for countries most impacted by climate change. It also called for an accelerated timeline for countries to revisit and strengthen their NDCs.

9. 2022: Loss and Damage Fund (COP27, Sharm El-Sheikh, Egypt)

- COP27 saw the historic establishment of a "Loss and Damage" fund, designed to support countries that are already facing severe climate-related impacts. Although the structure and funding sources are still being negotiated, this was a significant milestone in addressing climate justice for vulnerable nations. COP27 also reinforced the need to scale up climate adaptation and resilience-building efforts.

10. 2023: Enhanced Focus on Climate Finance and Adaptation (COP28, Dubai, UAE)

- COP28 continued the conversation on operationalizing the Loss and Damage fund and placed increased focus on financing mechanisms for both mitigation and adaptation. Discussions included sustainable energy transitions, with a focus on diversifying the funding sources for climate action. Adaptation financing and support for vulnerable countries were at the forefront, recognizing the need for financial pathways that support the resilience of least-developed nations.

Key Themes and Goals of COP29

COP29 is expected to build on past agreements and milestones, advancing concrete action on global climate commitments with an emphasis on urgency, inclusivity, and resilience. Here are the key themes and goals anticipated for COP29:

1. Strengthening Global Commitments to the 1.5°C Goal

- **Theme**: COP29 will reinforce the global objective to limit warming to 1.5°C above pre-industrial levels, as outlined in the Paris Agreement.

- **Goals**: Encourage countries to set and implement more ambitious Nationally Determined Contributions (NDCs). This includes urging nations to increase their short-term targets to achieve near-term emissions reductions, particularly by 2030, in line with scientific recommendations.

2. Operationalizing the Loss and Damage Fund

- **Theme**: Addressing climate-induced losses and damages in vulnerable countries, particularly for communities suffering from irreversible impacts like rising sea levels and extreme weather events.

- **Goals**: Finalize the governance, structure, and funding mechanisms for the Loss and Damage fund, including defining which countries and entities will contribute. The aim is to create a transparent and accessible process for distributing funds to affected countries and communities.

3. Scaling Up Climate Finance and Accessibility

- **Theme**: Mobilizing the financial resources needed to support climate action, with a particular focus on assisting developing and least-developed nations.

- **Goals**: Encourage developed nations to fulfill their commitments to provide at least $100 billion annually in climate finance. COP29 may also explore new sources of climate finance, including leveraging private sector investments, improving the accessibility of finance for adaptation projects, and expanding funding options for green technology and renewable energy.

4. Enhancing Adaptation and Resilience Building

- **Theme**: Helping countries build resilience to climate impacts, focusing on adaptation as a core pillar of global climate strategy.

- **Goals**: Establish specific metrics for tracking adaptation progress and increase funding dedicated to adaptation projects, such as sustainable infrastructure, disaster response systems, and nature-based solutions. COP29 may also work toward setting global adaptation targets, building upon the Paris Agreement's Global Goal on Adaptation.

5. Accelerating the Transition to Renewable Energy

- **Theme**: Reducing dependency on fossil fuels and scaling up investments in renewable energy sources like wind, solar, and hydroelectric power.

- **Goals**: Support commitments for phasing down coal and other high-emission energy sources. COP29 will likely address the role of clean energy finance, renewable infrastructure, and innovation to fast-track the global energy transition and align energy sector policies with net-zero objectives.

6. Advancing the Reduction of Methane Emissions

- **Theme**: Methane, a potent greenhouse gas, has emerged as a critical target for reducing short-term warming.

- **Goals**: Strengthen the Global Methane Pledge, with countries committing to reduce methane emissions by 30% by 2030 from 2020 levels. COP29 will likely seek greater international participation and may introduce monitoring mechanisms to track progress.

7. Promoting Climate Justice and Equity

- **Theme**: Recognizing and addressing the disproportionate impacts of climate change on vulnerable communities and emphasizing equitable solutions.

- **Goals**: Ensure that climate policies prioritize the needs of indigenous populations, small island nations, and economically disadvantaged communities. COP29 is expected to push for inclusivity in climate decision-making, promote community-led adaptation initiatives, and advance climate justice as a guiding principle in climate finance and policy.

8. Improving Transparency and Accountability in Climate Reporting

- **Theme**: Establishing clear and transparent reporting mechanisms for countries' climate actions, including emissions reductions and financial contributions.

- **Goals**: Enhance transparency frameworks for tracking NDC implementation, ensuring countries are accountable to their commitments. COP29 aims to further develop the "Global Stocktake" mechanism of the Paris Agreement, which will assess overall global progress on mitigation, adaptation, and finance.

9. Driving Technological Innovation and Knowledge Sharing

- **Theme**: Fostering new technologies and collaborative efforts to mitigate emissions, improve climate resilience, and accelerate the green transition.

- **Goals**: Promote public-private partnerships in green technology and renewable energy, encourage innovation in carbon capture, and support collaborative research efforts. COP29 may address ways to facilitate technology transfer to developing nations, especially for adaptation.

10. Enhancing Nature-Based Solutions and Ecosystem Restoration

- **Theme**: Recognizing the role of natural ecosystems in climate mitigation and adaptation, including carbon sequestration and biodiversity preservation.

- **Goals**: Support initiatives to restore forests, wetlands, and coastal ecosystems that can naturally capture carbon and protect biodiversity. COP29 will likely emphasize policies to combat deforestation, promote sustainable agriculture, and encourage conservation of marine and land ecosystems.

11. Building Youth Engagement and Inclusion

- **Theme**: Engaging the youth in climate action, acknowledging the growing influence of young activists and their demands for climate accountability.

- **Goals**: Amplify youth voices in negotiations, include youth representatives in decision-making processes, and provide platforms for youth-led climate solutions. COP29 will likely encourage countries to create pathways for youth to engage in climate projects and policy-making at national and international levels.

Chapter 2

The Road to COP29

Background on the Paris Agreement and COP28

The Paris Agreement, adopted at COP21 in 2015 in Paris, France, is a legally binding international treaty on climate change and represents a major milestone in global climate diplomacy. Its main goal is to limit global warming to well below 2°C above pre-industrial levels, with a more ambitious target of 1.5°C, to prevent the most severe impacts of climate change. Nearly every country in the world signed the agreement, making it a nearly universal pact.

Key Elements of the Paris Agreement

1. **Nationally Determined Contributions (NDCs)**: Countries submit their own climate action plans, known as NDCs, which outline their efforts to reduce emissions and adapt to climate impacts. The Paris Agreement requires countries to update their NDCs every five years, aiming for progressively ambitious targets.

2. **Global Goal on Adaptation**: Recognizing the urgent need for climate adaptation, the Paris Agreement introduced a collective goal to enhance countries' adaptive capacities, reduce vulnerability, and improve resilience to climate impacts.

3. **Financial Assistance and Climate Finance**: Developed countries committed to mobilizing at least $100 billion annually by 2020 to help developing nations with mitigation and adaptation efforts. This pledge also underpins the notion of climate justice, as many developing nations are among the most vulnerable to climate impacts despite contributing the least to global emissions.

4. **Transparency and Accountability**: The agreement established a transparency framework to ensure countries report their progress in a consistent manner. The first "Global Stocktake" (set for 2023) was designed to assess

collective progress towards the agreement's goals and encourage greater ambition.

5. **Long-Term Low-Emission Development Strategies**: Countries were encouraged to develop long-term strategies that would support their pathway to net-zero emissions by the second half of the century.

The Paris Agreement's flexible, inclusive framework allows each country to set its own climate goals while creating mechanisms to hold countries accountable, foster cooperation, and provide financial and technical support to those in need.

COP28 Overview

COP28, held in Dubai, UAE, in 2023, marked another important moment in the progression of the Paris Agreement and focused heavily on enhancing implementation, addressing financing gaps, and preparing for an intensified climate response. The conference was significant for several reasons:

Key Themes and Outcomes of COP28

1. **First Global Stocktake**: One of COP28's central features was the first Global Stocktake, which assessed the world's progress toward the Paris Agreement goals. The findings highlighted a significant gap between current emissions reductions and the pathway needed to limit global warming to 1.5°C. The stocktake underscored the need for more ambitious NDCs and accelerated action.

2. **Loss and Damage Fund Implementation**: COP28 continued discussions around operationalizing the Loss and Damage Fund established at COP27. This fund was created to provide financial support to countries and communities facing irreversible losses due to climate-related disasters, especially vulnerable countries that are on the frontline of climate impacts.

3. **Increased Focus on Climate Finance**: There was considerable emphasis on mobilizing and diversifying climate finance to support both mitigation and adaptation. COP28 built upon previous financial commitments, encouraging developed countries to meet the $100 billion

target, while also exploring private sector involvement, new financial tools, and funding channels.

4. **Phasing Down Fossil Fuels**: COP28 continued discussions on fossil fuel dependency, following COP26's "phase-down" language on coal and COP27's recommendations for accelerating the energy transition. Many stakeholders pushed for stronger commitments to reduce reliance on fossil fuels and increase investments in renewable energy infrastructure and clean technology.

5. **Adaptation and Resilience**: The conference strengthened the focus on adaptation, encouraging countries to prioritize resilience-building, especially for communities that are highly vulnerable to climate impacts. Initiatives included sustainable infrastructure development, early warning systems, and nature-based solutions like reforestation.

6. **Methane Reduction and Short-Lived Climate Pollutants**: Recognizing the immediate warming effects of methane, COP28 aimed to expand the Global Methane Pledge and encouraged countries to adopt methane reduction targets. This pledge, initiated in 2021, seeks to reduce global methane emissions by at least 30% from 2020 levels by 2030.

7. **Youth and Civil Society Engagement**: COP28 highlighted the voices of youth and civil society, acknowledging their role in pushing for greater climate ambition and accountability. Platforms were created to amplify youth-led climate solutions, and civil society representatives played an active role in negotiations, urging leaders to prioritize equity and climate justice.

8. **Advancements in Technology and Innovation**: COP28 underscored the importance of technological solutions, with a particular focus on clean energy technologies, carbon capture, and digital solutions for emissions tracking. Efforts to support technology transfer to developing countries also took center stage.

Significance of COP28 in Context of the Paris Agreement

COP28 served as a critical juncture, reaffirming the commitments of the Paris Agreement while highlighting areas where progress was lagging. By conducting the first Global Stocktake, COP28 provided a reality check on current global efforts and emphasized the need for urgent action, particularly for large emitters to enhance their NDCs.

Moreover, COP28 highlighted climate finance gaps and created renewed momentum for meeting the financial needs of developing countries. By expanding the Loss and Damage Fund and advancing discussions on adaptation, COP28 continued the work of the Paris Agreement to support vulnerable nations in facing the impacts of climate change.

The conference also reinforced the focus on climate justice, inclusivity, and accountability, ensuring that the Paris Agreement remains a dynamic framework capable of addressing the evolving challenges posed by climate change. COP28 set a clear path forward, underscoring the need for urgent, unified, and equitable action as the world prepares for COP29 and future milestones in global climate policy.

Major Global Climate Developments in the Past Year

The past year has seen significant climate-related developments across policy, finance, technology, and disaster response. These events have highlighted the urgent need for climate action and driven key global initiatives. Here are some of the major global climate developments leading up to COP29:

1. Global Stocktake at COP28

- One of the most critical developments of the past year was the Global Stocktake, the first comprehensive assessment of the world's progress under the Paris Agreement. Released at COP28, it highlighted a wide gap between current national commitments and the emission reductions needed to keep global warming within 1.5°C.

- The Stocktake underscored the urgency for countries to enhance their Nationally Determined Contributions (NDCs) and for a stronger push toward rapid decarbonization.

2. Operationalization of the Loss and Damage Fund

- The Loss and Damage Fund, initially agreed upon at COP27, moved closer to operational status in the past year. A coalition of developed and developing nations, along with non-governmental organizations, worked on defining the fund's structure, governance, and financing mechanisms.

- This fund is designed to provide financial support to vulnerable countries dealing with irreversible climate impacts, such as rising sea levels and extreme weather. Its development reflects increasing recognition of climate justice in international policy.

3. Record-Breaking Extreme Weather Events

- The past year saw numerous climate-induced disasters, from intense wildfires in Canada and Greece to devastating floods in Libya and Pakistan. Europe and North America also experienced unprecedented heatwaves, while countries like India and China faced both severe droughts and floods.

- These events underscored the need for immediate adaptation and resilience-building efforts. Governments and organizations called for accelerated investment in disaster response infrastructure and early warning systems, as the frequency and intensity of extreme weather events reached alarming levels.

4. Expansion of the Global Methane Pledge

- Methane, a potent greenhouse gas, has become a focal point in climate discussions due to its powerful short-term warming effect. Over the past year, additional countries signed on to the Global Methane Pledge, committing to cut methane emissions by at least 30% from 2020 levels by 2030.

- Initiatives to track and reduce methane emissions, particularly from oil and gas operations, agriculture, and waste, gained momentum. Technological advancements, like satellite monitoring of methane leaks, are helping to enforce accountability.

5. Surge in Climate Finance Commitments

- Financial institutions and governments committed unprecedented amounts of climate finance in the past year, with a strong focus on adaptation and resilience projects for vulnerable countries.

- Efforts to achieve the long-standing $100 billion per year climate finance target were intensified, with innovative financing mechanisms, including climate bonds, green funds, and public-private partnerships, designed to mobilize capital. Several major banks and investment funds also pledged to align their portfolios with net-zero goals.

6. Progress in Renewable Energy Deployment and Technology

- The past year saw record-breaking investments in renewable energy, particularly solar, wind, and battery storage. India, China, the EU, and the United States led the way in expanding their renewable energy capacities, while Africa and Latin America made significant strides in solar energy.

- Technological advances in energy storage, including scalable battery solutions and grid improvements, are making renewables more reliable and feasible as primary energy sources. Governments worldwide increasingly backed policies to accelerate the phase-out of coal and other fossil fuels, while renewable energy became more affordable.

7. Growing Focus on Climate Adaptation and Resilience

- Recognizing the limitations of mitigation efforts alone, governments placed greater emphasis on climate adaptation and resilience. In the past year, significant funding went toward adaptation projects in agriculture, water resources, and infrastructure.

- Organizations like the Green Climate Fund and the Adaptation Fund increased support for projects that build resilience in vulnerable regions. Initiatives include creating resilient crop systems, constructing climate-adaptive infrastructure, and expanding climate-smart agricultural practices.

8. Increased Corporate Commitments and Net-Zero Targets

- The corporate sector intensified its climate commitments, with many companies establishing Science-Based Targets (SBTs) aligned with net-zero pathways. Technology giants, global manufacturers, and financial institutions pledged to reduce emissions across their supply chains and operations.

- These net-zero goals have led to corporate investments in renewable energy, carbon offsetting projects, and innovations in low-carbon technologies. Some industries, especially aviation and shipping, have also explored alternative fuels, like sustainable aviation fuel and green hydrogen, as part of their decarbonization strategies.

9. Nature-Based Solutions and Ecosystem Restoration Initiatives

- Nature-based solutions (NbS), such as reforestation, wetland restoration, and sustainable agriculture, received heightened attention as essential strategies for both mitigation and adaptation. NbS projects were implemented worldwide, particularly in the Amazon, Southeast Asia, and Sub-Saharan Africa.

- Governments and organizations pushed for policies that incentivize ecosystem protection, reduce deforestation, and support regenerative agriculture, all of which are crucial for biodiversity preservation and carbon sequestration.

10. Youth and Civil Society Mobilization for Climate Justice

- Young activists and civil society organizations have increasingly influenced the global climate agenda, calling for stronger accountability and climate justice measures. In the past year, youth-led campaigns and grassroots organizations emphasized the need for inclusive and just climate policies.

- Movements advocating for climate justice highlighted the disproportionate impacts of climate change on marginalized communities and countries least responsible for emissions. Their advocacy has reinforced the importance of equitable policies, such as the Loss and Damage Fund, and has kept the spotlight on the responsibilities of developed nations.

11. Legislative and Policy Actions by Major Economies

- Major economies like the United States, European Union, China, and India introduced policies and incentives supporting the green transition. The EU advanced its "Fit for 55" package to reduce emissions by 55% by 2030, while the U.S. implemented the Inflation Reduction Act, incentivizing renewable energy and green technologies.

- These policies, along with similar measures in other regions, aim to reduce emissions across key sectors like energy, transportation, and industry, creating a pathway toward decarbonization and fostering green economic growth.

12. Advances in Climate-Related Science and Technology

- Over the past year, advancements in climate science and technology included improved climate models, which offer more accurate predictions of extreme weather events, and breakthroughs in carbon capture and storage technologies.

- Satellite data and artificial intelligence (AI) have enhanced tracking of emissions, deforestation, and climate impacts, making it easier to monitor progress toward climate goals. Carbon removal technologies, like direct air capture, have also shown potential as supplementary tools for mitigating emissions.

Preparations and Expectations from Major Stakeholders

As COP29 approaches, governments, NGOs, corporations, financial institutions, and civil society groups are ramping up their preparations and aligning their expectations. COP29 is anticipated to be a decisive moment for action, especially in light of the recent Global Stocktake showing that the world is not on track to meet the goals set by the Paris Agreement. Below is an overview of the preparations and expectations from each major stakeholder group.

1. Governments and Policymakers

Preparations:

- **Enhanced Nationally Determined Contributions (NDCs)**: Many governments are expected to revise their NDCs to reflect more ambitious emission reduction targets. There is also preparation for setting more transparent implementation and accountability measures.

- **Domestic Climate Policy Development**: Governments are working on legislation and policies to transition key sectors, including energy, transportation, and agriculture, to low-carbon pathways. Some are strengthening renewable energy incentives, carbon pricing, and policies to phase out fossil fuels.

- **Coalition Building**: Many countries are forming coalitions to address specific issues, such as the Fossil Fuel Non-Proliferation Treaty and the Beyond Oil and Gas Alliance, to collectively advocate for commitments on fossil fuel reductions and renewable energy transitions.

Expectations:

- **Commitments to Emission Reduction**: Governments are expected to adopt stronger commitments toward limiting global warming to 1.5°C, focusing on both short- and long-term goals.

- **Financial Support for Developing Countries**: Developing nations expect wealthier countries to deliver on climate finance commitments, particularly the $100 billion per year target, and contribute to the Loss and Damage Fund.

- **Global Agreement on Fossil Fuel Phase-Out**: Many governments, especially from vulnerable nations, are pushing for a formal commitment to phase down fossil fuel production and transition to clean energy by specific timelines.

2. Non-Governmental Organizations (NGOs) and Environmental Advocacy Groups

Preparations:

- **Research and Policy Proposals**: NGOs are preparing reports, policy briefs, and recommendations to influence negotiations at COP29. Many focus on areas like climate finance, loss and damage, adaptation, and equity.

- **Mobilizing Public Support**: Many NGOs are running campaigns and organizing events to raise public awareness and create a grassroots push for ambitious climate action. This includes mobilizing for rallies, petitions, and media campaigns leading up to COP29.

- **Advocacy for Climate Justice**: NGOs are preparing to highlight climate justice issues, emphasizing the disproportionate impact of climate change on marginalized and indigenous communities and the responsibility of major emitters.

Expectations:

- **Accountability for Previous Commitments**: NGOs expect governments and corporations to provide transparent reporting and accountability on previously made commitments, particularly regarding emissions, climate finance, and adaptation support.

- **Inclusion of Civil Society and Youth Voices**: There is a strong expectation for inclusivity at COP29, with NGOs advocating for meaningful involvement of civil society, youth representatives, and indigenous groups in discussions and decision-making.

- **Expansion of Loss and Damage Support**: NGOs hope to see concrete plans and substantial funding for the Loss and Damage Fund, focusing on addressing the needs of communities most affected by climate disasters.

3. Corporate Sector and Private Industry

Preparations:

- **Setting and Aligning Net-Zero Targets**: Companies are increasingly committing to net-zero targets and are preparing to announce new or updated plans aligned with Science-Based Targets (SBTs) during COP29. They are investing in renewable energy, supply chain decarbonization, and low-carbon technologies.

- **Investment in Green Technology and Innovation**: Many corporations are directing investments toward green technology, carbon capture, sustainable agriculture, and alternative fuels, anticipating both regulatory pressures and market demands for sustainability.

- **Increased Reporting and Transparency**: Companies are enhancing their ESG (Environmental, Social, and Governance) reporting frameworks to provide greater transparency on their climate-related actions, emissions, and progress towards net-zero.

Expectations:

- **Clear Policy Signals and Incentives**: The corporate sector expects clear and stable policy frameworks to support the low-carbon transition, including carbon pricing, incentives for green investment, and policies on renewable energy.

- **Public-Private Collaboration**: There is an expectation for increased collaboration with governments and international bodies to develop scalable solutions, particularly in areas like carbon markets, sustainable infrastructure, and technological innovation.

- **Guidance on Carbon Markets and Offsetting**: Many companies are looking for clearer standards and guidelines on carbon credits and offset markets to ensure credible offsets and avoid greenwashing.

4. Financial Institutions and Investors

Preparations:

- **Commitments to Climate Finance and Sustainable Investment**: Financial institutions are preparing announcements on sustainable finance commitments, including investments in renewable energy, green bonds, and funds directed towards climate adaptation and resilience.

- **Development of Climate Risk Assessment Tools**: Banks and investors are advancing climate risk assessment tools to evaluate the exposure of their portfolios to climate risks, ensuring they align with the goals of the Paris Agreement.

- **Engagement with Regulatory Bodies**: Financial institutions are collaborating with regulators and policymakers to establish clear frameworks for sustainable finance, including green taxonomies and climate disclosure requirements.

Expectations:

- **Scaling Up Climate Finance**: There is an expectation for governments and multilateral organizations to increase climate finance commitments, particularly for adaptation and resilience, to mobilize private capital effectively.

- **Enhanced Climate Disclosure Requirements**: Investors seek consistent climate disclosure standards to enable accurate assessment of companies' climate-related risks and alignment with climate targets.

- **Clear Taxonomies for Sustainable Finance**: Financial stakeholders expect COP29 to advance the development of a global green finance taxonomy that defines what qualifies as a sustainable investment to prevent "greenwashing."

5. Youth and Civil Society

Preparations:

- **Youth-Led Campaigns and Advocacy Initiatives**: Youth climate groups are preparing campaigns, protests, and social media initiatives to amplify their calls for urgent action and highlight the intergenerational impacts of climate change.

- **Representation and Participation**: Youth representatives are working to ensure their participation in COP29 discussions, particularly around topics like climate justice, education, and sustainable futures.

- **Proposals for Just Transition**: Many civil society groups are focused on promoting a "just transition" that addresses economic and social inequities, ensuring that climate action benefits all communities fairly and does not disproportionately impact vulnerable groups.

Expectations:

- **Commitment to 1.5°C Goal and Ambitious Action**: Young people and civil society activists are pushing for governments to make concrete commitments that will keep global warming below 1.5°C.

- **Focus on Climate Justice and Equity**: There is a strong expectation that COP29 will address climate justice, ensuring that climate action is equitable and that the voices of marginalized communities are represented and respected.

- **Inclusion of Youth in Decision-Making**: Youth groups expect a more inclusive process with meaningful opportunities to influence decisions, particularly as they represent the generation most impacted by the future climate scenario.

6. Indigenous and Local Communities

Preparations:

- **Advocacy for Land Rights and Ecosystem Protection**: Indigenous communities are working to raise awareness of the critical role they play in conservation, pushing for the protection of their land rights and advocating for ecosystem-based approaches to climate adaptation.

- **Documentation of Climate Impacts and Solutions**: Indigenous groups are preparing to share case studies on the impacts of climate change on their communities and the traditional knowledge they bring to climate resilience and biodiversity protection.

- **Coalitions and Alliances for Climate Justice**: Indigenous groups are collaborating with NGOs and climate justice organizations to strengthen their representation and advocate for policies that respect indigenous rights and knowledge.

Expectations:

- **Recognition of Indigenous Knowledge and Rights**: Indigenous communities seek formal recognition of their knowledge systems and expect COP29 to support policies that uphold their rights and prioritize the protection of their lands.

- **Support for Nature-Based Solutions**: There is a strong push for nature-based solutions that align with indigenous approaches to land stewardship and biodiversity conservation.

- **Focus on Climate Resilience and Adaptation**: Indigenous groups hope COP29 will increase support for adaptation and resilience measures that reflect the specific needs and priorities of their communities.

Chapter 3

Key Players and Negotiating Parties

List of Countries, Organizations, And Alliances at COP29

Here is a comprehensive list of countries, organizations, and alliances expected to participate in COP29. This list reflects the broad coalition of governments, international organizations, non-governmental organizations, alliances, and civil society groups working toward global climate action:

1. Countries and Regional Coalitions

- **Developed Countries:**
    - United States, Canada, European Union (EU member states like Germany, France, Spain, Italy), United Kingdom, Australia, New Zealand, Japan, South Korea.

- **Developing Countries:**
    - India, China, Brazil, South Africa, Indonesia, Mexico, Nigeria, Kenya, Ethiopia.

- **Small Island Developing States (SIDS):**
    - Maldives, Fiji, Barbados, Tuvalu, Seychelles, Marshall Islands.

- **Least Developed Countries (LDCs):**
    - Afghanistan, Bangladesh, Nepal, Uganda, Mozambique, and many other vulnerable nations facing disproportionate climate impacts.

- **Regional Coalitions:**
    - **African Group of Negotiators (AGN):** Represents African nations in climate negotiations, with a focus on adaptation, climate finance, and technology transfer.

- Alliance of Small Island States (AOSIS): A coalition of small island countries advocating for strong mitigation to prevent sea-level rise and extreme weather.

- Climate Vulnerable Forum (CVF): Comprises countries highly susceptible to climate impacts, emphasizing the need for ambitious climate action and adaptation support.

- Like-Minded Developing Countries (LMDCs): Includes countries like India, China, and Saudi Arabia, often emphasizing economic development needs in climate discussions.

2. International Organizations

- United Nations (UN):

  - United Nations Framework Convention on Climate Change (UNFCCC): Hosts the COP and provides administrative and logistical support.

  - United Nations Development Programme (UNDP): Works on sustainable development and climate resilience projects, especially in developing countries.

  - United Nations Environment Programme (UNEP): Focuses on global environmental issues, providing research and advocacy on climate action.

- World Bank Group: Offers climate finance, including funding for adaptation and resilience in developing countries.

- International Monetary Fund (IMF): Involved in policy support for climate-related financial risks and green economic growth.

- International Renewable Energy Agency (IRENA): Promotes global adoption of renewable energy, supporting countries with transition pathways.

- **World Health Organization (WHO)**: Engages in climate-health discussions, given the link between climate change and public health challenges.

- **Food and Agriculture Organization (FAO)**: Works on climate-smart agriculture and food security in the face of climate change.

- **Green Climate Fund (GCF)**: Major source of climate finance, particularly for adaptation projects in vulnerable countries.

3. Climate Alliances and Coalitions

- **The High Ambition Coalition (HAC)**: A coalition of countries that advocate for ambitious climate targets, aiming to limit global warming to 1.5°C.

- **Powering Past Coal Alliance (PPCA)**: Comprising countries and organizations committed to phasing out coal power.

- **Global Methane Pledge**: An alliance committed to reducing methane emissions by 30% by 2030, with members like the U.S., EU, and over 100 other countries.

- **Beyond Oil and Gas Alliance (BOGA)**: Initiated by Denmark and Costa Rica, it includes countries committed to phasing out oil and gas production.

- **Glasgow Financial Alliance for Net Zero (GFANZ)**: A coalition of private financial institutions working to accelerate the transition to a net-zero global economy.

- **C40 Cities Climate Leadership Group**: A network of megacities worldwide dedicated to implementing climate action at the city level.

- **Under2 Coalition**: An alliance of subnational governments committed to reducing greenhouse gas emissions, including states, provinces, and cities.

- **Indigenous Peoples Organizations (IPOs)**: Groups advocating for indigenous rights and climate justice, often in coalition with other NGOs and grassroots organizations.

4. Non-Governmental Organizations (NGOs) and Environmental Advocacy Groups

- **Climate Action Network (CAN)**: A global network of over 1,500 NGOs working on climate and energy issues, with regional hubs on every continent.

- **Greenpeace International**: A major environmental advocacy group that mobilizes public opinion and action against climate change.

- **World Wildlife Fund (WWF)**: Focuses on conservation and advocates for sustainable environmental practices and climate resilience.

- **Friends of the Earth (FoE)**: International organization advocating for environmental justice and social change to combat climate impacts.

- **Environmental Defense Fund (EDF)**: Works on policy and market-based solutions to reduce greenhouse gas emissions.

- **The Nature Conservancy (TNC)**: Focuses on nature-based solutions and ecosystem preservation to mitigate climate impacts.

- **350.org**: A climate action organization that mobilizes global campaigns, focusing on reducing fossil fuel reliance and promoting clean energy.

5. Financial and Business Institutions

- **Glasgow Financial Alliance for Net Zero (GFANZ)**: A group of financial institutions committed to aligning their portfolios with net-zero emissions.

- **Net-Zero Asset Owner Alliance**: A coalition of major asset owners committing to carbon-neutral investment portfolios.

- **Principles for Responsible Investment (PRI)**: An organization supporting investors in considering ESG factors, including climate risks, in their portfolios.

- **Banking and Investment Giants**: Banks such as Bank of America, BNP Paribas, HSBC, and BlackRock are expected to participate and discuss their roles in sustainable finance and green investments.

- **Corporate Climate Leaders**: Companies with significant climate action initiatives, including tech giants like Google, Microsoft, and Apple, as well as sectors like renewable energy, electric vehicle manufacturers, and sustainable fashion.

6. Research Institutions and Think Tanks

- **Intergovernmental Panel on Climate Change (IPCC)**: Provides scientific reports that inform COP negotiations, particularly regarding climate impacts, vulnerabilities, and adaptation.

- **International Institute for Environment and Development (IIED)**: Focuses on policy research related to climate resilience and sustainable development.

- **World Resources Institute (WRI)**: Provides data and analysis on global climate, sustainability, and urban development.

- **Stockholm Environment Institute (SEI)**: A research organization working on issues such as climate adaptation, sustainable cities, and climate policy.

- **Center for Climate and Energy Solutions (C2ES)**: Works on practical climate solutions, offering research and policy guidance for businesses and policymakers.

7. Youth and Civil Society Movements

- **Fridays for Future**: A youth-led movement inspired by Greta Thunberg, demanding ambitious climate action from global leaders.

- **Youth Climate Leaders**: Networks and alliances of young climate advocates from various countries who actively participate in COP discussions.

- **Civil Society and Climate Justice Coalitions**: Includes groups like Extinction Rebellion, Global Witness, and various indigenous rights organizations that advocate for climate equity and accountability.

- **Indigenous Youth Networks**: Indigenous youth coalitions that focus on climate resilience, land rights, and protecting biodiversity, highlighting the unique vulnerabilities faced by indigenous communities.

8. Technology and Innovation Alliances

- **Mission Innovation**: An alliance focused on accelerating clean energy innovation through global collaboration among governments and industry.

- **Clean Energy Ministerial (CEM)**: A global forum promoting policies and programs that advance clean energy technology.

- **Breakthrough Energy**: Founded by Bill Gates, this initiative invests in technologies to accelerate the green transition, including clean energy, carbon capture, and sustainable agriculture solutions.

- **Global Alliance for Smart Cities**: Supports climate-smart urban infrastructure, promoting the role of technology in reducing cities' carbon footprints.

The Role of Major Economies

Major economies such as the U.S., the European Union (EU), China, and India play pivotal roles in COP negotiations, climate commitments, and global climate action due to their significant economic influence and contributions to global emissions. Here's an in-depth look at their roles and expectations:

1. United States

- **Role and Influence**: As one of the world's largest greenhouse gas emitters and a key economic power, the U.S. has considerable sway in COP negotiations. The U.S. has historically played both supportive and ambivalent roles in global climate action, with its commitment fluctuating with changes in administration. Under the Biden administration, the U.S. has aimed to reassert its leadership on climate issues.

- **Commitments and Actions**:

    o **Emission Targets**: The U.S. has pledged to cut its greenhouse gas emissions by 50-52% below 2005 levels by 2030, aligning with the Paris Agreement's goals.

    o **Domestic Policies**: Key legislation, like the Inflation Reduction Act (IRA), includes substantial investments in clean energy, carbon reduction, and climate resilience.

    o **International Climate Finance**: The U.S. has committed to increasing climate finance to support developing countries in achieving their climate goals, though it has faced challenges in meeting its financial pledges.

- **Key Areas of Focus at COP29**:

    o Scaling up climate finance for developing nations.

    o Strengthening partnerships to accelerate renewable energy development and technology sharing.

- Promoting initiatives to reduce methane emissions, including through the Global Methane Pledge, which it co-leads with the EU.

2. European Union

- **Role and Influence**: The EU has long been a leader in climate diplomacy and policy, often pushing for ambitious climate targets and working collaboratively within its member states to address climate issues. With one of the world's most comprehensive climate action frameworks, the EU is influential in encouraging other countries to adopt stricter climate policies.

- **Commitments and Actions**:
    - **Emission Targets**: The EU aims to achieve a 55% reduction in greenhouse gas emissions from 1990 levels by 2030, with a legally binding commitment to reach net-zero emissions by 2050.

    - **Domestic Policies**: Initiatives like the European Green Deal and "Fit for 55" package include measures to transform energy systems, promote clean industries, and enhance biodiversity protection.

    - **Climate Finance Leadership**: The EU remains a top provider of climate finance, supporting developing countries in areas like renewable energy, sustainable agriculture, and resilience building.

- **Key Areas of Focus at COP29**:
    - Advancing the global adoption of carbon pricing mechanisms and carbon border adjustments.

    - Enhancing global commitments to a just transition, emphasizing equity and social welfare.

    - Pushing for an international framework to phase out fossil fuels, specifically coal.

3. China

- **Role and Influence**: China, as the world's largest emitter and a major economic power, plays a crucial role in climate negotiations. China has positioned itself as a leader in renewable energy investment, while also emphasizing its need for economic growth and development as it gradually reduces coal dependence.

- **Commitments and Actions**:

  - **Emission Targets**: China has committed to peak its carbon emissions before 2030 and achieve carbon neutrality by 2060.

  - **Domestic Policies**: China is a global leader in solar and wind power production, electric vehicles, and battery storage technology. It has invested heavily in renewable infrastructure, though coal still remains a significant part of its energy mix.

  - **South-South Climate Cooperation**: China provides climate finance and technical assistance to other developing countries, promoting green development and sustainable projects.

- **Key Areas of Focus at COP29**:

  - Advocating for differentiated responsibilities, emphasizing that developed countries should bear a larger share of the climate burden.

  - Supporting initiatives for loss and damage compensation for vulnerable countries.

  - Enhancing global collaboration on renewable energy technologies, as well as green finance.

4. India

- **Role and Influence**: As a major emerging economy and one of the world's largest emitters, India balances its commitment to climate action with its developmental needs. India has been a strong voice for climate equity, often advocating that developed countries provide more support to developing nations.

- Commitments and Actions:

  o Emission Targets: India has pledged to achieve net-zero emissions by 2070 and aims to meet 50% of its energy requirements from renewable sources by 2030.

  o Domestic Policies: Initiatives like the National Solar Mission, Pradhan Mantri Kisan Urja Suraksha (focused on solar pumps for agriculture), and its leadership of the International Solar Alliance highlight India's focus on renewable energy.

  o Climate Justice and Equity: India consistently emphasizes the need for climate justice, arguing that developed countries should take greater responsibility for their historical emissions and support developing nations.

- Key Areas of Focus at COP29:

  o Advocating for increased climate finance and technology transfer to support developing countries.

  o Pushing for international support in addressing "loss and damage" from climate impacts.

  o Strengthening commitments to renewable energy expansion and adaptation measures that benefit developing countries.

Perspectives From Vulnerable Nations and Island States

At COP29, vulnerable nations and island states bring essential perspectives to the table, emphasizing the disproportionate impacts of climate change they face and their need for urgent, tailored support. These nations, which often contribute the least to global emissions, are on the frontlines of rising sea levels, extreme weather, and ecosystem degradation. Their priorities at COP29 reflect a strong call for ambitious climate action from major emitters and equitable access to resources for adaptation and resilience.

1. Emphasis on Urgent Climate Action and Adaptation

- **Immediate Threats**: Many island states and vulnerable nations are already experiencing severe impacts, including coastal erosion, saltwater intrusion, and increased frequency of hurricanes, droughts, and floods. Nations like the Maldives, Tuvalu, and Fiji face existential threats from rising sea levels, which could displace entire communities and, in extreme cases, render some areas uninhabitable.

- **Adaptation Support**: These countries require significant support for adaptation measures such as sea walls, flood defenses, and climate-resilient infrastructure. However, the cost of such measures is often beyond their reach without external assistance.

- **Demand for Clear Roadmaps and Implementation**: Vulnerable nations call for major economies to provide clear, actionable plans and timelines for achieving emissions reductions and to fund adaptation projects effectively.

2. Loss and Damage Compensation

- **Climate Justice and Accountability**: The idea of loss and damage revolves around compensating countries that suffer irreversible climate impacts, often due to emissions from industrialized nations. Vulnerable countries argue that historical and high-emission nations should be accountable for the climate harm endured by those least responsible.

- **Dedicated Funding Mechanism**: Countries like Vanuatu, Bangladesh, and the Marshall Islands are advocating for the establishment of a formal and dedicated loss and damage fund. They want this fund to provide financial support directly to communities facing immediate and long-term losses, such as those from extreme weather events or slow-onset processes like desertification.

- **Operationalizing Loss and Damage from COP28**: At COP28, there were steps taken to establish a loss and damage fund. Vulnerable nations at COP29 seek to make the fund operational with clear guidelines on how it will be funded, disbursed, and governed, ensuring it is readily accessible and avoids bureaucratic delays.

3. Equitable Climate Finance and Accessibility

- **More Inclusive Funding**: Vulnerable nations argue that traditional climate finance often does not reach those most in need. The complexity of accessing funds, coupled with requirements that smaller and less-developed nations may not meet, limits their ability to benefit from climate finance.

- **Calls for Grant-Based Finance**: These countries are advocating for more grant-based finance rather than loans, as loans can exacerbate their debt burdens, limiting their economic growth and resilience. They emphasize the need for a fair distribution of climate funds, particularly for adaptation and resilience projects.

- **Increasing Climate Finance Commitments**: Many vulnerable nations highlight that developed countries have yet to fulfill previous climate finance commitments, such as the pledge to mobilize $100 billion annually for developing countries. They seek assurances that promises made at COP29 will be honored promptly.

4. Advocacy for Just Transition and Climate Equity

- **Principle of Common but Differentiated Responsibilities (CBDR)**: Vulnerable nations and island states consistently advocate for the CBDR principle, which recognizes that while all countries have a role in combating climate change, developed nations bear greater responsibility due to their historical emissions.

- **Support for Economic and Social Transition**: Many developing countries are highly dependent on fossil fuels or carbon-intensive industries and need help transitioning to green economies. They argue for international support that prioritizes poverty reduction, energy access, and economic stability as part of a "just transition" that is inclusive and fair.

- **Global South Solidarity**: Many island nations have forged alliances with other vulnerable countries, forming blocs like the Alliance of Small Island States (AOSIS) and the Climate Vulnerable Forum (CVF), which work collectively to emphasize that climate policies must address the needs of those most affected by climate change.

5. Advancing Nature-Based Solutions and Sustainable Development

- **Focus on Ecosystem-Based Adaptation**: Vulnerable nations frequently promote nature-based solutions that align with their local ecosystems and cultural practices. Examples include mangrove restoration for coastal protection, sustainable agriculture for food security, and ecosystem conservation to support biodiversity.

- **Sustainable Development Pathways**: These nations argue that climate resilience must be integrated with sustainable development goals, including poverty reduction, health care access, and education. Many vulnerable nations see climate action as an opportunity to drive sustainable, low-carbon growth.

- **Protecting Biodiversity and Marine Resources**: Island states are especially vocal about the need to protect ocean ecosystems, which are critical to their economies and cultural identities. They advocate for stronger global commitments to protect marine biodiversity, address overfishing, and reduce pollution from plastic and other sources.

6. Technology Transfer and Capacity Building

- **Access to Green Technologies**: Vulnerable nations argue for equitable access to technology that can help them adapt to and mitigate climate impacts. They seek increased support for renewable energy infrastructure, climate-resilient agriculture, early warning systems, and water management technologies.

- **Capacity Building and Knowledge Sharing**: Many vulnerable nations lack the resources and expertise to implement advanced climate solutions. They emphasize the importance of capacity-building programs and technical support, including knowledge-sharing partnerships with more developed countries and organizations.

- **Digital Innovation for Resilience**: Some nations are advocating for more digital resources to track climate impacts, improve resilience, and enhance transparency. Technological support also enables remote and island nations to monitor critical climate data in real-time, aiding in early warning and response.

7. Raising Global Awareness and Climate Advocacy

- **Amplifying the Voices of the Vulnerable**: Vulnerable nations emphasize the need for their experiences and stories to be heard on the global stage. They have been increasingly active in ensuring that larger nations and global audiences recognize the human cost of climate inaction.

- **Youth and Indigenous Voices**: Many island nations bring youth activists and indigenous representatives to COP29 to speak about the cultural, economic, and personal impacts of climate change. These advocates provide firsthand accounts and emphasize that preserving their heritage and ways of life depends on urgent climate action.

- **Moral Appeal for Global Solidarity**: Vulnerable nations frequently make a moral appeal for global solidarity, stressing that climate change is not only a scientific or economic issue but a matter of justice, ethics, and human rights. They urge wealthier nations to acknowledge the ethical imperative to support those facing the brunt of climate impacts.

Chapter 4

Climate Science and Updates

Latest IPCC Findings

The latest findings from the Intergovernmental Panel on Climate Change (IPCC) reinforce the urgency for comprehensive and immediate action to combat climate change, warning that the world is at a critical juncture. The IPCC's most recent synthesis report, part of its Sixth Assessment Cycle (AR6), compiles years of scientific research and builds on three previous reports covering the physical science of climate change, its impacts and adaptation potential, and mitigation pathways. Here are some of the key findings:

1. Global Warming and Temperature Rise

- **Warming Trends**: The IPCC reports that human activities have caused global temperatures to rise by approximately 1.1°C above pre-industrial levels, with temperatures likely to exceed 1.5°C between 2030 and 2035 if current trends continue.

- **Impacts of 1.5°C vs. 2°C**: Limiting warming to 1.5°C will reduce risks significantly compared to 2°C. Beyond 1.5°C, the likelihood of extreme weather events, biodiversity loss, and irreversible damage to ecosystems increases dramatically.

- **Irreversible Changes**: Some impacts, like the melting of polar ice caps and rising sea levels, may become irreversible within the century, leading to lasting consequences for coastal communities and ecosystems.

2. Impacts on Ecosystems and Biodiversity

- **Ecosystem Stress**: Climate change is pushing many ecosystems to the brink, with increased temperatures, ocean acidification, and extreme weather events leading to severe biodiversity losses. Coral reefs, for instance, are expected to decline by 70-90% at 1.5°C and face near-total loss at 2°C.

- **Species Extinction Risks**: Many species face higher extinction risks, particularly in fragile ecosystems like the Arctic, alpine regions, and tropical forests. Disruption of food chains and habitats is also contributing to ecosystem collapse in some areas.

- **Human Impact on Natural Resources**: Deforestation, land-use change, and pollution are intensifying climate impacts, compounding the degradation of ecosystems and exacerbating biodiversity loss.

3. Extreme Weather Events and Human Vulnerability

- **Increase in Extreme Events**: The frequency and intensity of extreme weather events—including heatwaves, floods, droughts, and hurricanes—are increasing. The IPCC highlights that these events are more severe than previously predicted and are directly linked to human-induced climate change.

- **Health and Livelihoods**: The direct impact on human health, agriculture, and water security is significant. Heatwaves and wildfires, for instance, are leading to increased mortality, especially in vulnerable populations. Crop yields are declining due to extreme weather, impacting food security.

- **Economic Costs**: The economic toll from climate-related disasters is rising, with vulnerable regions bearing the brunt. The report stresses that without effective adaptation, the economic damage will exacerbate inequality and lead to higher costs over time.

4. Adaptation and Resilience

- **Need for Immediate Action**: While adaptation efforts are underway, current measures are insufficient. Many countries, particularly in the Global South, lack the resources and technology to implement adequate adaptation strategies.

- **Localized Solutions**: Adaptation must be context-specific and consider local socio-economic and environmental conditions. The IPCC suggests that nature-based solutions—

like restoring mangroves and forests—can offer cost-effective resilience benefits, but these alone are insufficient without broader climate action.

- **Limits to Adaptation**: The IPCC highlights that some impacts cannot be adapted to, emphasizing the need to keep warming below critical thresholds to prevent catastrophic losses.

5. Greenhouse Gas Emissions and Carbon Budgets

- **Global Emissions Trends**: Greenhouse gas emissions continue to rise despite increased awareness and policy efforts, with energy production and industrial activities as the primary contributors. Current policies are insufficient to meet the targets set by the Paris Agreement.

- **Carbon Budget Constraints**: To limit warming to 1.5°C, the IPCC estimates that the remaining carbon budget will be exhausted within this decade at the current rate of emissions. Immediate and drastic reductions in emissions are necessary to stay within this limit.

- **Importance of Decarbonization**: The report stresses the need to accelerate the transition from fossil fuels to renewable energy, electrify transportation and industry, and implement carbon capture and storage technologies where feasible.

6. Mitigation Pathways and Technology

- **Rapid Transition to Clean Energy**: The IPCC states that scaling up renewable energy sources (such as wind, solar, and hydropower) is crucial. Decarbonizing the energy sector would account for a significant portion of the emission reductions needed.

- **Innovative Technologies and Practices**: Emerging technologies like green hydrogen, carbon capture and storage (CCS), and energy-efficient building materials are highlighted as key to meeting emission reduction targets. However, these technologies need rapid scaling and investment.

- **Sustainable Land Management**: Forest conservation, reforestation, and sustainable agriculture practices are essential to reduce land-based emissions. The IPCC suggests that changes in land use could contribute up to 20% of the required mitigation by mid-century.

7. Social Equity and Climate Justice

- **Disproportionate Impacts on Vulnerable Communities**: The findings underscore that marginalized populations, including low-income communities, Indigenous peoples, and small island states, are disproportionately affected by climate change. These communities often have less capacity to adapt and fewer resources to recover.

- **Need for Inclusive Policies**: Climate policies must address social equity to ensure that vulnerable populations receive the support they need. The IPCC calls for just transition strategies that prioritize workers and communities reliant on fossil fuel industries to prevent economic displacement.

- **Climate Finance and Responsibility**: Wealthier nations are urged to meet their commitments to provide financial assistance to developing nations. The IPCC stresses that substantial investment is needed to both mitigate and adapt to climate impacts, and that climate finance is crucial to a global response.

8. Urgency for Global Cooperation and Policy Integration

- **International Collaboration**: Effective climate action requires unprecedented cooperation between countries, especially between developed and developing nations. The IPCC suggests that a global framework for transparent reporting, financing, and accountability is essential.

- **Integrated Policy Approach**: Addressing climate change cannot be isolated from other policy areas like public health, economic planning, and infrastructure development. The IPCC advises that climate considerations should be integrated into all levels of policy-making to ensure comprehensive and cohesive responses.

- **The Need for Immediate Action**: With only a narrow window left to avert the worst outcomes, the IPCC emphasizes the urgency of bold policy shifts and collective global action within this decade. It warns that delays will increase the likelihood of severe and irreversible impacts, making mitigation and adaptation efforts more challenging and costly.

New Research on Global Warming, Sea Level Rise, And Biodiversity Loss

Recent research on global warming, sea level rise, and biodiversity loss highlights the acceleration of climate impacts, emphasizing that the interconnected consequences are progressing faster than previously anticipated. These studies draw attention to more precise measurements, projections, and new discoveries in how climate change is affecting ecosystems, communities, and natural systems globally. Here are some key developments and insights from the latest research:

1. Global Warming Trends and Temperature Extremes

- **Record Temperatures**: Studies show that the past decade was the warmest on record, with temperatures rising faster than expected. This year has seen unprecedented temperature spikes globally, especially in the Northern Hemisphere, which scientists attribute to both human-induced climate change and periodic natural factors like El Niño events.

- **Feedback Loops**: Research reveals that warming is triggering feedback mechanisms, such as permafrost thaw and changes in ocean circulation, which release more greenhouse gases (like methane and CO_2) into the atmosphere. This accelerates warming, creating a self-reinforcing cycle that is difficult to reverse.

- **Projected Warming Impacts**: New models indicate that, if current emissions trends continue, global temperatures could rise by 3°C or more by 2100. The implications are severe, with such warming likely leading to increasingly destructive weather patterns, agriculture failures, and ecosystem collapses.

2. Sea Level Rise and Coastal Impacts

- **Acceleration of Sea Level Rise**: The rate of sea level rise has doubled over the past two decades, with projections indicating that sea levels could increase by up to 2 meters by 2100 if emissions remain high. Melting ice sheets in

Greenland and Antarctica are significant contributors, with research showing that these ice losses are occurring faster than anticipated.

- **Risk to Coastal Cities and Small Island Nations**: Studies estimate that coastal flooding could affect up to 630 million people by mid-century. Cities like Miami, Dhaka, and Tokyo, as well as entire island nations, face the prospect of increased flooding, saltwater intrusion, and displacement.

- **Tipping Points in Ice Melt**: Recent research highlights that certain parts of the West Antarctic Ice Sheet may have reached irreversible melting points. This would result in long-term sea level rise even if emissions were cut significantly, pointing to the importance of rapid adaptation and resilience planning for coastal regions.

3. Biodiversity Loss and Ecosystem Disruption

- **Species Extinction Rates**: Biodiversity loss is accelerating, with extinction rates now estimated to be 1,000 times higher than pre-industrial levels. Habitat loss, climate change, pollution, and over-exploitation are driving this trend, which threatens a million species with extinction in the coming decades.

- **Climate-Induced Range Shifts**: New findings indicate that many species are migrating to higher altitudes or latitudes in search of cooler climates. However, the rate of these shifts is often insufficient to keep up with warming, leaving many species vulnerable to local extinction as their habitats become uninhabitable.

- **Loss of Key Ecosystems**: Coral reefs, rainforests, and Arctic habitats are among the most threatened ecosystems. For instance, coral reefs are projected to decline by 70-90% at 1.5°C of warming, and almost entirely at 2°C, affecting marine biodiversity, coastal protection, and food resources for millions of people.

4. Impacts on Marine and Terrestrial Life

- **Ocean Warming and Acidification**: The ocean absorbs about 90% of excess heat from global warming, leading to marine heatwaves that devastate ecosystems like coral reefs and kelp forests. Additionally, increased CO_2 absorption is causing ocean acidification, which harms shell-forming organisms (e.g., coral, mollusks), affecting the entire marine food web.

- **Forest Ecosystems and Wildfires**: Forests, which store large amounts of carbon, are increasingly affected by wildfires, pests, and disease—all exacerbated by climate change. Research indicates that prolonged droughts and higher temperatures are leading to more frequent and intense wildfires, transforming some forests from carbon sinks to carbon sources.

- **Pollinator Decline**: Pollinator species, critical to global food production, are declining due to habitat loss, pesticide exposure, and climate impacts. This trend poses a risk to food security, with recent studies emphasizing that warming is affecting pollinator behavior and disrupting plant-pollinator relationships.

5. Human Health and Socioeconomic Impacts

- **Health Risks from Extreme Heat**: Rising temperatures are associated with increased mortality and morbidity, especially during heatwaves. Research indicates that heat-related deaths are expected to rise as climate change intensifies, affecting vulnerable populations the most.

- **Food and Water Security Threats**: Climate change is reducing agricultural yields by affecting crop growth conditions, particularly in already warm regions. Changing precipitation patterns and prolonged droughts are also depleting freshwater sources, impacting millions who depend on glacier-fed rivers and rain-fed agriculture.

- **Migration and Displacement**: Sea level rise, extreme weather, and resource scarcity are forcing communities to migrate. Studies predict that up to 200 million people could be displaced by climate impacts by 2050, with many becoming "climate refugees."

6. Technological and Policy Responses

- **Renewable Energy Advancements**: Research highlights that transitioning to renewable energy sources could play a key role in limiting warming to 1.5°C. Solar, wind, and battery technology advancements are making renewables more accessible, but scaling them rapidly enough to meet targets remains a challenge.

- **Conservation and Restoration**: Ecosystem conservation and restoration are seen as critical to mitigating climate impacts. Reforestation, wetland restoration, and coastal ecosystem preservation can enhance biodiversity, store carbon, and improve resilience to climate impacts.

- **Adaptation Policies**: The research emphasizes that while mitigation remains crucial, adaptation policies—such as building flood defenses, adopting climate-resilient agriculture, and improving early warning systems—are essential for minimizing the impacts already locked in by existing emissions.

Innovations In Climate Modeling and Forecasting

Recent innovations in climate modeling and forecasting are enhancing our understanding of climate change, enabling more precise predictions of its impacts, and improving strategies for adaptation and mitigation. As computing power and data accessibility expand, scientists are leveraging these advancements to develop models that are not only more accurate but also more regionally focused, interdisciplinary, and capable of representing complex feedback mechanisms within the climate system. Here are some of the most significant innovations in climate modeling and forecasting:

1. High-Resolution Climate Models

- **Global and Regional Precision**: Traditional climate models operated at coarse resolutions, often 100 km or more, limiting their accuracy in capturing small-scale phenomena. New high-resolution models can simulate down to a few kilometers, allowing for more detailed projections of local climate impacts, such as urban heatwaves, flash floods, and coastal erosion.

- **Cloud and Storm Simulation**: Higher resolution improves the representation of clouds, storms, and convection patterns, which are essential for accurately predicting extreme weather events. This detail is crucial for forecasting localized impacts, particularly in regions vulnerable to severe storms and hurricanes.

2. Earth System Models (ESMs) with Complex Feedbacks

- **Inclusion of Biosphere-Atmosphere Interactions**: Modern ESMs now integrate not just the atmosphere and oceans but also interactions with the biosphere, such as vegetation, soil carbon, and oceanic ecosystems. This integration helps to predict how changes in one component, like deforestation, can feedback into the climate system, affecting carbon storage and temperatures.

- **Carbon and Methane Cycles**: Many ESMs now simulate carbon and methane feedbacks from permafrost, forests, and oceans, which are critical for understanding long-term warming. These cycles influence the amount of greenhouse gases in the atmosphere, providing a more holistic view of emissions sources and sinks.

3. Artificial Intelligence and Machine Learning

- **Data-Driven Climate Prediction**: Machine learning models are being used to process large datasets, identify patterns, and make predictions with greater speed and accuracy. For example, AI algorithms can analyze satellite data to improve cloud and precipitation modeling, enhance short-term weather forecasting, and reduce uncertainties in climate predictions.

- **Speed and Efficiency**: Machine learning enables faster model runs, allowing researchers to test a broader range of scenarios and conduct ensemble simulations that produce probabilistic climate forecasts. AI also improves efficiency, as it can optimize complex calculations and provide real-time updates with less computing power.

4. Cloud-Based Computing and Big Data Analytics

- **Global Climate Data Sharing**: Cloud computing platforms allow researchers worldwide to share, analyze, and model large datasets collaboratively. This approach accelerates model development and enables the integration of multiple sources of data, such as satellite imagery, ocean temperature records, and emissions data.

- **Handling Large Ensembles**: Big data processing capabilities make it possible to run "ensemble" models that generate multiple simulations under varied conditions, improving the robustness of climate forecasts. These ensembles can reveal likely outcomes under different emissions scenarios and are valuable for policymakers in risk assessment and planning.

5. Regional Climate Models for Localized Forecasting

- **Tailoring for Specific Regions**: Regional Climate Models (RCMs) provide more localized insights by downscaling global climate data for specific areas, such as Europe, Southeast Asia, or the Arctic. These models are critical for local adaptation efforts, allowing communities to prepare for specific risks like flooding, droughts, and heatwaves.

- **Urban and Coastal Applications**: Urban-specific climate models simulate how cities amplify heat (urban heat islands), while coastal models focus on sea-level rise, erosion, and storm surges. These region-specific models provide actionable insights for urban planners, policymakers, and infrastructure developers.

6. Integrated Assessment Models (IAMs) for Policy Analysis

- **Linking Climate and Socio-Economic Outcomes**: IAMs combine physical climate models with economic, social, and environmental data, allowing scientists to assess the impacts of various policy choices. By linking emissions pathways with social and economic outcomes, IAMs help in understanding the trade-offs between different climate policies, like carbon pricing or renewable energy adoption.

- **Long-Term Scenarios**: IAMs support scenario analysis, such as projecting the global economy's response to different emissions reduction pathways. This helps governments and organizations weigh the long-term costs and benefits of mitigation strategies and plan for a low-carbon transition.

7. Interactive and Dynamic Climate Visualization

- **Real-Time Climate Dashboards**: Interactive platforms now provide real-time climate information and forecasts, accessible to scientists, policymakers, and the public. These dashboards visualize climate projections, heat maps, and sea-level rise data, making complex information more comprehensible for a broader audience.

- **Educational and Decision-Making Tools**: Visual tools offer simulations where users can adjust variables like emissions levels or land use changes to see their impact on temperature, precipitation, or extreme weather. This functionality enhances climate education and enables more informed decision-making at all levels.

8. Emulator Models for Fast Projections

- **Simplified, High-Speed Simulations**: Climate emulators are simplified models that provide rapid projections of climate outcomes, useful when full model runs would be too time-intensive. Emulators are invaluable for testing the effects of different policies and emission pathways quickly, offering guidance for time-sensitive decisions.

- **Policy and Scenario Testing**: Emulators can simulate "what-if" scenarios to evaluate the impact of actions like reforestation, carbon taxes, or renewable energy adoption. These models can approximate outcomes, facilitating a faster response to emerging climate challenges.

9. Enhanced Data from Remote Sensing and Satellite Observations

- **Satellite Monitoring of Greenhouse Gases**: Satellites now provide real-time monitoring of CO_2, methane, and other greenhouse gases across the globe. This data improves model accuracy and helps track emission hotspots, such as industrial sites or deforested areas, allowing for quicker regulatory responses.

- **Dynamic Climate Observations**: Satellite data on sea ice thickness, ocean temperatures, land cover, and forest health provide continuous updates, enhancing the accuracy of predictions related to sea-level rise, drought patterns, and biodiversity impacts.

10. Climate Sensitivity and Tipping Points

- **Refined Estimates of Climate Sensitivity**: New research is refining estimates of how sensitive Earth's climate is to CO_2 increases. Understanding this "climate sensitivity" helps predict how much warming is likely under different emissions scenarios, refining projections for key policy thresholds, like 1.5°C or 2°C of warming.

- **Tipping Point Detection**: Advanced models are working to identify and quantify tipping points—irreversible changes, such as Arctic ice melt or Amazon rainforest dieback. Improved detection of tipping points allows for early warning and emphasizes the urgency of action to avoid crossing these critical thresholds.

Chapter 5

Adaptation and Resilience

Strategies Discussed to Help Nations Adapt to Climate Change Impacts

Adapting to climate change is increasingly essential as its impacts - rising temperatures, extreme weather, sea level rise, and biodiversity loss - become more pronounced. Nations are focusing on both immediate and long-term adaptation strategies, particularly to protect vulnerable populations, strengthen infrastructure, and sustain ecosystems. Here are some of the major adaptation strategies being discussed and implemented at a national and international level:

1. Climate-Resilient Infrastructure Development

- **Flood-Resistant Infrastructure**: In flood-prone areas, many countries are building levees, flood barriers, and improved drainage systems to handle extreme rainfall and storm surges. For example, cities like Rotterdam have invested in advanced water management and flood protection systems.

- **Elevated and Reinforced Buildings**: Coastal and riverine communities are constructing buildings on stilts, using water-resistant materials, and retrofitting existing structures to withstand stronger storms, hurricanes, and floods.

- **Green Infrastructure**: Incorporating nature-based solutions such as green roofs, permeable pavements, and urban parks helps reduce the urban heat island effect, manage stormwater, and improve air quality.

2. Ecosystem-Based Adaptation (EbA)

- **Mangrove Restoration for Coastal Protection**: Many coastal nations are restoring and protecting mangroves to act as natural barriers against storm surges and erosion. Mangroves also absorb carbon, benefiting both adaptation and mitigation efforts.

- **Forest Conservation and Reforestation**: By preserving natural forests and reforesting degraded lands, countries can improve local climates, prevent soil erosion, and enhance biodiversity. Forests also provide essential ecosystem services and support livelihoods for communities that depend on them.

- **Wetland Restoration**: Wetlands act as natural buffers that absorb floodwaters and prevent coastal erosion. They are also rich in biodiversity and act as significant carbon sinks, making their restoration a priority in many regions.

3. Climate-Smart Agriculture and Food Security

- **Drought-Resistant Crops**: Farmers in arid and semi-arid regions are adopting drought-resistant crop varieties to ensure productivity even during periods of low rainfall. Genetically modified or selectively bred crops are being developed to withstand extreme weather.

- **Sustainable Water Management**: Nations are implementing irrigation systems that use less water, such as drip irrigation and rainwater harvesting, to make agriculture more resilient to water scarcity. These methods improve water efficiency and reduce reliance on unpredictable rainfall.

- **Diversification of Crop Varieties**: Growing a variety of crops reduces the risk of total crop failure due to extreme weather or pests. Diversification also promotes soil health and long-term agricultural resilience.

4. Disaster Preparedness and Early Warning Systems

- **Advanced Forecasting Technologies**: Investments in satellite monitoring, meteorological stations, and data-sharing networks enhance early warning systems for extreme weather events, giving communities time to prepare and evacuate when necessary.

- **Community-Based Early Warning Systems**: Localized warning systems that integrate local knowledge and technological tools (like mobile alerts) help reach remote and vulnerable populations quickly, enhancing community preparedness.

- **Emergency Response Training**: Training communities in disaster response, first aid, and evacuation procedures prepares them to act quickly during emergencies, reducing injury and loss of life. For example, nations prone to cyclones like Bangladesh have trained thousands in emergency preparedness and evacuation procedures.

5. Water Resource Management and Conservation

- **Sustainable Groundwater Management**: Over-extraction of groundwater is a concern in many countries, especially with drought and reduced rainfall. Improved regulation, artificial recharge methods, and conservation strategies help sustain groundwater supplies.

- **Desalination and Water Recycling**: Water-scarce countries like Israel and Saudi Arabia are investing in desalination and wastewater recycling to ensure a reliable supply of fresh water. These technologies are crucial for regions facing chronic water shortages.

- **Integrated Water Management**: By implementing integrated approaches to manage rivers, lakes, and watersheds, countries can prevent conflicts over water resources, improve water quality, and better handle seasonal fluctuations in water availability.

6. Urban Adaptation and Resilience

- **Urban Greening Initiatives**: Cities are planting trees, creating green belts, and establishing urban forests to counteract urban heat, reduce pollution, and provide spaces for stormwater absorption. Initiatives like New York City's Million Trees project are examples of large-scale urban greening.

- **Cooling Centers and Heat Action Plans**: Cities facing extreme heat are establishing cooling centers and creating heat action plans that include early warnings, public awareness campaigns, and support for vulnerable populations like the elderly.

- **Climate-Sensitive Urban Planning**: Integrating climate adaptation into zoning laws, building codes, and city planning ensures that new developments are designed with climate resilience in mind. This includes reducing impervious surfaces and planning infrastructure that can withstand extreme weather events.

7. Insurance and Financial Protection Mechanisms

- **Climate Risk Insurance**: Nations are promoting insurance products that protect against climate-related risks, such as crop insurance for farmers and flood insurance for homeowners. This financial safety net helps people recover and rebuild after disasters.

- **Climate Bonds and Resilience Funds**: Governments and international organizations are issuing climate bonds and creating resilience funds to finance adaptation projects, such as constructing resilient infrastructure and supporting disaster relief efforts.

- **Parametric Insurance for Quick Payouts**: Some countries are adopting parametric insurance, which disburses funds quickly based on pre-determined metrics (e.g., rainfall levels or wind speed) to support rapid disaster recovery.

8. Health Adaptation Strategies

- **Heat Health Action Plans**: As heat waves become more frequent, many countries are implementing action plans to prevent heat-related illnesses, particularly in vulnerable populations. This includes creating public awareness campaigns, establishing cooling centers, and providing medical training for heat-related emergencies.

- **Vector Control Programs**: With climate change influencing the spread of diseases like malaria and dengue, nations are investing in vector control programs to prevent outbreaks. These include mosquito control initiatives, public health education, and early detection systems.

- **Strengthening Healthcare Systems**: Enhancing healthcare infrastructure, especially in rural and underserved regions, prepares systems to handle climate-induced health risks. Training healthcare providers on the specific challenges posed by climate change, like new infectious diseases, is also a priority.

9. Migration and Relocation Planning

- **Managed Retreat**: In areas where adaptation is no longer feasible, governments are planning for managed retreat, which involves relocating communities away from high-risk areas, such as low-lying islands or flood-prone regions.

- **Protection for Climate Migrants**: Nations and international bodies are discussing legal protections for climate migrants—people forced to leave their homes due to climate impacts. Initiatives include creating migration pathways and support systems for those displaced by climate change.

- **Economic Development in Safe Areas**: Developing economic opportunities in safer inland or elevated regions can incentivize migration away from high-risk coastal and flood-prone areas.

10. International Cooperation and Funding for Adaptation

- **Global Climate Funds**: The Green Climate Fund (GCF) and Adaptation Fund are key sources of finance to help developing countries implement adaptation projects. These funds support efforts in the Global South, where resources for adaptation are often limited.

- **Technology Transfer and Capacity Building**: Wealthier countries are investing in technology transfer and capacity building in developing nations, allowing them to implement climate adaptation strategies and improve local resilience. This includes sharing knowledge, technology, and best practices.

- **Cross-Border Water and Resource Management**: International agreements help countries manage shared resources, such as river basins and forests, to prevent conflicts and improve resilience in transboundary ecosystems.

Examples Of Adaptation Programs and Funding

Adaptation programs and funding initiatives are critical in helping communities and ecosystems cope with the impacts of climate change. Here are some notable examples of adaptation programs and the funding mechanisms that support them:

1. Green Climate Fund (GCF)

- **Overview**: Established in 2010 under the United Nations Framework Convention on Climate Change (UNFCCC), the GCF is one of the largest international funds dedicated to climate adaptation and mitigation.

- **Funding Mechanism**: The GCF supports projects in developing countries to improve climate resilience, reduce emissions, and protect vulnerable communities. Contributions come from developed nations, and as of now, billions have been committed.

- **Example Projects**:

    - *Ecosystem-Based Adaptation in Colombia*: This project focuses on preserving ecosystems like wetlands and forests to reduce climate impacts on agriculture and water supply.

    - *Climate-Resilient Infrastructure in Bangladesh*: The GCF is funding projects to improve flood defenses and ensure safe drinking water access, which benefits communities in flood-prone coastal regions.

2. Adaptation Fund (AF)

- **Overview**: The Adaptation Fund, established under the Kyoto Protocol, provides financial resources for adaptation projects and programs in developing countries. It emphasizes projects that directly support vulnerable communities.

- **Funding Mechanism**: The AF is financed through a share of proceeds from certified emissions reductions (CERs) issued under the UN's Clean Development Mechanism (CDM), along with donor contributions.

- **Example Projects**:
  - *Agriculture Resilience in Senegal.* This project enhances farmers' resilience to droughts through improved water management and climate-smart farming practices.

  - *Community-Based Flood Preparedness in Cambodia.* The fund supports training for local communities in disaster response and sustainable water management practices.

3. Pilot Program for Climate Resilience (PPCR)

- **Overview**: A part of the Climate Investment Funds (CIF), the PPCR focuses on mainstreaming climate resilience in development planning and financing climate-resilient infrastructure and agriculture.

- **Funding Mechanism**: The PPCR provides grants and concessional funding to developing countries, allowing them to invest in resilience projects across sectors.

- **Example Projects**:
  - *Climate-Resilient Roads in Nepal.* Investments help develop roads that can withstand landslides and extreme weather, ensuring connectivity for rural communities.

  - *Resilient Crop Varieties in Mozambique.* The PPCR supports the adoption of drought-tolerant crops and sustainable farming practices to secure food supplies in the face of erratic rainfall.

4. Least Developed Countries Fund (LDCF)

- **Overview**: Managed by the Global Environment Facility (GEF), the LDCF provides funding specifically to the world's least developed countries (LDCs) to address urgent adaptation needs.

- **Funding Mechanism**: The LDCF receives voluntary contributions from developed countries, and it supports projects identified as priorities by the LDCs themselves.

- **Example Projects**:

    o *Climate-Resilient Water Supplies in Malawi:* The project aims to improve water supply systems to withstand droughts and floods, securing clean water access for rural areas.

    o *Sustainable Land Management in Sudan:* This initiative supports soil and water conservation techniques to adapt agricultural systems to the country's increasing aridity.

5. European Union's Adaptation Initiatives

- **Overview**: The European Union (EU) funds adaptation projects both within Europe and in developing countries through programs like the LIFE program and the European Climate Adaptation Platform.

- **Funding Mechanism**: EU funds for adaptation come from several sources, including the LIFE Climate Action program and the European Regional Development Fund (ERDF), which supports member countries in strengthening climate resilience.

- **Example Projects**:

    o *Urban Heat Adaptation in Southern Europe:* Funding supports urban greening projects, public cooling centers, and heat alert systems in cities across southern Europe to reduce heatwave impacts.

    o *Flood Risk Reduction in the Netherlands:* The EU contributes to projects that develop natural flood defenses, such as rewilding rivers and enhancing coastal wetlands.

6. Climate Adaptation and Mitigation Program for Aral Sea Basin (CAMP4ASB)

- **Overview**: Funded by the World Bank, this program addresses the water scarcity crisis in Central Asia by enhancing water and land management practices to adapt to regional climate impacts.

- **Funding Mechanism**: The program is financed through a combination of World Bank loans, grants, and contributions from Central Asian countries.

- **Example Projects**:

  o *Irrigation Efficiency in Uzbekistan and Kazakhstan*: Investments in modern irrigation systems and improved water storage help reduce water waste and secure resources for agriculture in arid areas.

  o *Drought-Resistant Crops and Farmer Training*: Farmers are provided with drought-resistant seeds and trained in climate-smart practices to enhance resilience in agricultural production.

7. Africa Adaptation Initiative (AAI)

- **Overview**: Launched by African nations, the AAI focuses on strengthening climate resilience across Africa, with a focus on adaptation financing, capacity building, and disaster risk reduction.

- **Funding Mechanism**: AAI collaborates with the African Development Bank (AfDB) and international donors to pool resources for adaptation projects continent-wide.

- **Example Projects**:

  o *Early Warning Systems in East Africa*: Funding goes toward advanced weather monitoring and community-based early warning systems to protect vulnerable communities from extreme weather events.

- o *Climate-Smart Agriculture in Sub-Saharan Africa.* Programs funded by AAI promote sustainable farming practices that conserve water and soil, supporting food security amid erratic rainfall.

8. International Climate Initiative (IKI)

- **Overview**: IKI, launched by the German government, funds climate adaptation and biodiversity conservation projects worldwide, especially in developing countries.

- **Funding Mechanism**: The initiative is funded by Germany's federal budget and provides both grants and loans for a variety of adaptation and conservation projects.

- **Example Projects**:

  - o *Protecting Coastal Ecosystems in Southeast Asia.* IKI funds mangrove restoration projects to protect against storm surges and enhance biodiversity in coastal areas.

  - o *Community-Based Adaptation in Pacific Islands.* Projects focus on empowering island communities to protect their resources and adapt to rising sea levels through innovative coastal management practices.

9. Japan's Joint Crediting Mechanism (JCM)

- **Overview**: Japan's JCM provides financial and technical assistance to developing countries for climate adaptation, while enabling Japan to count the emissions reductions toward its own targets.

- **Funding Mechanism**: Japanese public and private sector entities finance projects abroad, focusing on both adaptation and mitigation.

- **Example Projects**:

  - o *Energy-Efficient Water Management in Southeast Asia.* Projects improve water management infrastructure to make it more resilient to climate change while reducing energy use.

    o *Disaster-Resilient Infrastructure in the Pacific Islands:* Japan funds climate-resilient building materials and designs for communities vulnerable to typhoons and rising sea levels.

10. Private Sector and Philanthropy Initiatives

- **Insurance Companies Offering Climate-Resilience Policies**: Firms like AXA and Swiss Re are developing climate-resilient insurance products for individuals, businesses, and governments. These policies cover damages from extreme weather and provide financial security.

- **Philanthropic Foundations (e.g., Rockefeller Foundation, Gates Foundation)**: Many philanthropic organizations fund local adaptation projects and innovative climate resilience solutions, including early warning systems and health adaptation programs.

Focus On Agriculture, Water Resources, And Urban Resilience

Addressing climate adaptation in agriculture, water resources, and urban resilience has become a top priority as climate impacts intensify globally. Here are examples of key strategies, programs, and funding mechanisms focused on these areas:

1. Agriculture Adaptation Programs

Agriculture faces profound challenges from climate change, including shifts in growing seasons, extreme weather, and water scarcity. Adaptation programs in agriculture focus on improving resilience through sustainable practices, climate-smart farming, and resilient infrastructure.

Notable Programs and Funding

- **Climate-Smart Agriculture (CSA):**
    - **Overview**: CSA combines sustainable agricultural practices with improved resource management to reduce greenhouse gas emissions and adapt to climate impacts.
    - **Funding Mechanism**: Programs like the World Bank's CSA initiative and the FAO support farmers with technical assistance, funding, and training in climate-smart practices.
    - **Examples**:
        - *Resilient Crop Varieties*: CSA programs promote drought-resistant and heat-tolerant crop varieties, enabling farmers in arid regions to maintain productivity despite erratic rainfall.
        - *Agroforestry in Kenya*: Combining trees with crops to improve soil fertility and water retention, this approach helps shield farms from climate extremes and fosters biodiversity.

- International Fund for Agricultural Development (IFAD):
  - Overview: IFAD supports smallholder farmers in developing countries to strengthen resilience through climate-smart technologies, training, and financial resources.
  - Funding Mechanism: The Adaptation for Smallholder Agriculture Programme (ASAP) is an IFAD initiative that integrates climate resilience in rural agricultural projects.
  - Examples:
    - *Drip Irrigation in Morocco:* By supporting drip irrigation, IFAD helps farmers optimize water use, improving crop yields while conserving limited water resources.
    - *Climate Resilient Farmer Cooperatives in Latin America:* Cooperatives receive funds for climate-resilient crop storage and improved market access, helping farmers adapt to uncertain weather patterns.

2. Water Resource Adaptation Programs

As climate change exacerbates droughts and water shortages, particularly in arid and semi-arid regions, water resource management programs focus on efficient use, storage, and equitable distribution of water.

Notable Programs and Funding

- Green Climate Fund (GCF):
  - Overview: GCF's portfolio includes multiple water resource projects that address water scarcity, improve infrastructure, and enhance storage in vulnerable regions.
  - Funding Mechanism: Grants and concessional financing are available to fund sustainable water resource management.

- o Examples:
    - *Integrated Water Resource Management in Jordan:* GCF funds rainwater harvesting and desalination plants, increasing freshwater availability and resilience to drought.

    - *Flood Management in the Philippines:* Funding supports construction of resilient infrastructure like flood barriers and water channels to protect against seasonal monsoons.

- **UNESCO's Water Resilience Programs:**
  - o **Overview:** UNESCO focuses on building the capacity of countries to manage water resources sustainably through education, policy support, and technology.

  - o **Examples:**
    - *Groundwater Management in Africa:* UNESCO collaborates with African nations to develop sustainable groundwater systems, safeguarding access to clean water even during droughts.

    - *Community-Based Water Resilience in Central Asia:* UNESCO supports local water user associations to manage scarce water supplies, ensuring equitable access and reducing conflict over resources.

3. Urban Resilience Programs

Cities are increasingly vulnerable to extreme weather, heatwaves, and flooding. Urban resilience programs focus on strengthening infrastructure, green spaces, and emergency preparedness to protect people and assets in cities.

Notable Programs and Funding

- 100 Resilient Cities (100RC):
  - Overview: Founded by the Rockefeller Foundation, 100RC assists cities in building capacity for urban resilience, including climate adaptation, disaster response, and risk management.
  - Funding Mechanism: The initiative provides funding, technical support, and access to a global network of experts.
  - Examples:
    - *Heat Action Plans in Ahmedabad, India*: To combat rising urban temperatures, 100RC assists in creating cooling centers, public outreach campaigns, and emergency response plans for vulnerable populations.
    - *Flood Prevention in New Orleans, USA*: The city uses green infrastructure (e.g., rain gardens and permeable pavements) to manage stormwater and reduce flood risks.

- Global Covenant of Mayors for Climate & Energy:
  - Overview: This coalition of cities worldwide commits to climate action and resilience, helping urban areas reduce emissions and adapt to climate risks.
  - Funding Mechanism: The initiative is funded through contributions from cities, philanthropic organizations, and public grants.
  - Examples:
    - *Green Roofs and Vertical Gardens in Paris*: Initiatives to create more green spaces reduce the urban heat island effect, improve air quality, and enhance biodiversity.

- *Coastal Resilience in Miami, USA*: Funding supports storm surge barriers, wetland restoration, and resilient building codes to protect the city from rising sea levels.

- **European Bank for Reconstruction and Development (EBRD)**:

  o **Overview**: EBRD supports climate resilience in cities by financing infrastructure upgrades, energy efficiency projects, and green spaces.

  o **Examples**:

    - *Climate-Resilient Public Transport in Sarajevo*: By investing in flood-proof public transport systems, EBRD helps Sarajevo's transit infrastructure remain operational during extreme weather.

    - *Stormwater Management in Belgrade*: EBRD funds sustainable drainage systems to control stormwater, reduce urban flooding, and protect residential and commercial areas.

Chapter 6

Mitigation Commitments and Emissions Targets

Nationally Determined Contributions (NDCs) Updates

Nationally Determined Contributions (NDCs) are central to achieving the targets set by the Paris Agreement, which aims to limit global warming to well below 2°C, and ideally 1.5°C, above pre-industrial levels. NDCs represent each country's commitments to reduce greenhouse gas (GHG) emissions and adapt to climate impacts. They are updated periodically, typically every five years, allowing countries to raise their ambitions over time as technology advances, climate impacts become more evident, and global expectations grow.

Here's an overview of recent updates and key aspects in NDCs as of COP29:

1. Enhanced Ambition in Emission Reduction Targets

- **More Stringent Targets**: Many countries have responded to the IPCC's recent findings and global pressure by updating their NDCs with more ambitious emission reduction targets. Countries including the EU member states, the U.S., Japan, and the UK have committed to significant cuts by 2030.

- **Examples of Updated Commitments**:

   o *United States*: Increased its 2030 target to reduce emissions by 50-52% below 2005 levels.

   o *European Union*: Raised its target to reduce emissions by at least 55% by 2030 from 1990 levels.

   o *Japan*: Announced plans to cut emissions by 46% by 2030, an increase from its previous 26% goal, signaling greater ambition.

2. Focus on Adaptation and Resilience

- **Adaptation Plans**: Many updated NDCs now include adaptation strategies to address the increased risk of climate impacts like heatwaves, floods, and sea-level rise. Countries are setting concrete goals to protect vulnerable sectors such as agriculture, water resources, and coastal infrastructure.

- **Examples**:

    o *Bangladesh*: Committed to protecting agriculture and water resources through infrastructure upgrades and climate-smart practices.

    o *Fiji*: Focused on coastal resilience to address rising sea levels, with specific targets for flood defense and relocation plans for vulnerable communities.

3. Incorporation of Nature-Based Solutions

- **Ecosystem Protection**: Nature-based solutions, such as reforestation, wetland restoration, and sustainable land management, are increasingly included as part of countries' NDCs. These approaches help to sequester carbon while providing resilience against climate impacts.

- **Examples**:

    o *Brazil*: Recommitted to reducing deforestation rates in the Amazon and enhancing forest protections to achieve net-zero deforestation by 2030.

    o *Colombia*: Included targets for restoring degraded lands and protecting ecosystems as part of its broader climate strategy.

4. Support for Renewable Energy Transition

- **Energy Transformation Goals**: Countries are setting specific targets for expanding renewable energy, such as wind, solar, and hydropower, as part of their NDC updates. Many are committing to phasing out coal and other fossil fuels.

- **Examples**:

    o *India*: Announced a target of achieving 50% of its electricity from renewable sources by 2030 and reducing emissions intensity by 45% from 2005 levels.

o *South Korea:* Plans to increase its renewable energy share to 30% by 2030, focusing on solar and offshore wind projects.

5. Increased Financial and Technical Support

- **Climate Finance Commitments**: Wealthier nations have pledged increased financial support for adaptation and mitigation in developing countries. This support helps bridge the gap for countries that face significant barriers to achieving their climate targets.

- **Examples**:

  o *European Union:* Raised its financial contributions to developing countries to support green infrastructure, clean energy, and adaptation projects.

  o *Canada and Germany:* Increased their commitments to the Green Climate Fund (GCF) to aid vulnerable countries in enhancing their NDCs.

6. New Carbon Pricing Mechanisms and Market-Based Approaches

- **Carbon Markets and Trading**: Some countries are now including or expanding carbon pricing and emissions trading systems (ETS) in their NDCs. These mechanisms encourage reductions by putting a price on carbon emissions.

- **Examples**:

  o *China:* Recently launched a nationwide emissions trading system (ETS), which covers power generation and incentivizes companies to reduce emissions.

  o *Canada:* Expanded its carbon tax and committed to progressively increasing the carbon price over the next decade to reach emissions reduction targets.

7. Commitments to Methane Reduction

- **Methane Emissions**: Methane has emerged as a significant focus in recent NDC updates, as it is a potent greenhouse gas with a high short-term impact on warming. Many countries have pledged to reduce methane emissions from oil, gas, agriculture, and waste sectors.

- Examples:
    - *United States*: Pledged a 30% reduction in methane emissions by 2030 in line with the Global Methane Pledge.

    - *Argentina and Nigeria*: Committed to targeting methane reductions in the agricultural sector, focusing on livestock management and waste reduction practices.

8. Enhanced Transparency and Accountability Mechanisms

- **Tracking and Reporting**: Many countries have strengthened transparency mechanisms within their NDCs, improving data collection and reporting frameworks. This allows for better monitoring of progress and encourages accountability on climate commitments.

- Examples:
    - *Chile and South Africa*: Integrated improved reporting frameworks into their NDCs, aligning with international standards for transparent emissions tracking.

    - *EU*: Enhanced its climate monitoring system to ensure real-time tracking of member states' progress toward climate goals.

Prominent Emissions Reduction Pledges

Several countries and regions have made ambitious emissions reduction pledges in recent years, emphasizing the need for strong climate action to meet the goals of the Paris Agreement. Here are some prominent examples:

1. United States

- **Target:** Reduce greenhouse gas emissions by 50-52% below 2005 levels by 2030.

- **Details:** This target aligns with the Biden administration's commitment to net-zero emissions by 2050. Major policy initiatives include expanding renewable energy, electrifying transportation, and incentivizing industries to reduce emissions through clean energy investments and federal standards.

- **Key Initiatives:** Inflation Reduction Act (IRA), aiming to accelerate renewable energy projects, support electric vehicles, and invest in clean manufacturing.

2. European Union

- **Target:** Reduce emissions by at least 55% below 1990 levels by 2030.

- **Details:** The EU's European Green Deal and its "Fit for 55" legislative package outline the pathway to achieve these goals, which include expanding renewable energy, improving energy efficiency, and adopting sustainable agriculture and industrial practices.

- **Key Initiatives:** Implementation of a stricter Emissions Trading System (ETS), boosting renewable energy infrastructure, and establishing a Carbon Border Adjustment Mechanism (CBAM) to prevent carbon leakage.

3. China

- **Target:** Peak carbon dioxide emissions before 2030 and achieve carbon neutrality by 2060.

- **Details**: As the world's largest emitter, China's pledge is significant, aiming for a rapid transition from coal to renewable energy sources. China is investing heavily in solar, wind, and electric vehicle technology, while gradually reducing its dependence on coal.

- **Key Initiatives**: A national Emissions Trading Scheme (ETS), the world's largest, covers the power sector initially but is expected to expand to other sectors in coming years.

4. India

- **Target**: Reduce emissions intensity by 45% from 2005 levels and achieve 50% cumulative electric power capacity from non-fossil fuel-based energy sources by 2030.

- **Details**: India has focused on increasing renewable energy, particularly solar power, to achieve these targets. India also set a goal for net-zero emissions by 2070, with a strong emphasis on clean energy and energy efficiency measures.

- **Key Initiatives**: Expanding solar power infrastructure under initiatives like the International Solar Alliance (ISA) and ambitious plans to increase renewable energy capacity to 500 GW by 2030.

5. Japan

- **Target**: Reduce emissions by 46% from 2013 levels by 2030 and aim for carbon neutrality by 2050.

- **Details**: Japan has set policies to increase energy efficiency, adopt hydrogen as a clean energy source, and reduce coal power reliance. The government is supporting innovations in hydrogen, energy storage, and nuclear technology.

- **Key Initiatives**: The Green Growth Strategy, focusing on decarbonization of industries like steel, chemicals, and transportation, and supporting clean energy innovation and investment.

6. United Kingdom

- **Target**: Reduce emissions by 68% below 1990 levels by 2030 and 78% by 2035.

- **Details**: The UK has made one of the most ambitious short-term reduction pledges among developed nations. Key strategies include phasing out coal power, transitioning to renewable energy sources, and banning the sale of new petrol and diesel vehicles by 2030.

- **Key Initiatives**: The Ten Point Plan for a Green Industrial Revolution, which includes offshore wind, hydrogen production, green public transport, and sustainable agriculture.

7. South Korea

- **Target**: Reduce emissions by 40% below 2018 levels by 2030 and achieve carbon neutrality by 2050.

- **Details**: South Korea is focusing on renewable energy, hydrogen, and electric vehicles, while reducing dependence on coal. The country has also pledged to phase out overseas coal financing.

- **Key Initiatives**: The Korean Green New Deal, focusing on green infrastructure, renewable energy, and electric vehicle incentives, and expansion of the South Korean ETS.

8. Brazil

- **Target**: Reduce emissions by 50% below 2005 levels by 2030, with a focus on stopping illegal deforestation in the Amazon and achieving net-zero by 2050.

- **Details**: Brazil's pledge includes a significant focus on preserving the Amazon rainforest, which plays a crucial role in carbon sequestration. Policies aim to combat illegal deforestation, expand renewable energy, and promote sustainable agriculture.

- **Key Initiatives**: Investment in biofuels, policies for forest conservation, and expansion of clean energy sources like hydroelectric, wind, and solar.

9. Canada

- **Target**: Reduce emissions by 40-45% below 2005 levels by 2030, with a long-term goal of net-zero by 2050.

- **Details**: Canada's strategy focuses on reducing oil and gas sector emissions, carbon pricing, expanding renewable energy, and implementing clean fuel standards. The Canadian government is also investing in nature-based solutions for carbon sequestration.

- **Key Initiatives**: Strengthening carbon pricing, promoting electric vehicles, and implementing the Clean Fuel Standard to reduce lifecycle emissions of fuels.

Carbon Capture and Offset Mechanisms Discussed

At COP29, carbon capture and offset mechanisms are expected to be key topics of discussion as countries seek ways to reduce their greenhouse gas (GHG) emissions and meet their climate targets. These mechanisms are seen as critical tools for achieving net-zero emissions goals, especially as some sectors (such as heavy industry and aviation) may struggle to decarbonize fully in the near term.

1. Carbon Capture and Storage (CCS)

Carbon Capture and Storage (CCS) involves capturing carbon dioxide (CO_2) emissions from sources like power plants and industrial processes before they are released into the atmosphere, and then storing them underground in geological formations. CCS has become an important part of many countries' decarbonization strategies, especially for sectors where emissions are difficult to eliminate.

- **Key Strategies**:
    - **Capture**: CO_2 is separated from exhaust gases using various technologies, such as amine-based absorption, membrane filtration, or cryogenic distillation.
    - **Transport**: Captured CO_2 is then transported via pipelines or ships to storage sites.
    - **Storage**: CO_2 is injected into deep underground rock formations, such as depleted oil and gas reservoirs or deep saline aquifers, where it is intended to remain permanently.

- **Current Challenges**:
    - **Cost**: CCS is expensive, with high initial investment and ongoing operational costs. As a result, it often requires government subsidies or high carbon prices to be financially viable.
    - **Scale**: Scaling CCS technologies to a level that can have a significant impact on global emissions

remains a significant challenge. Current projects capture a relatively small portion of global CO_2 emissions.

- o **Public Acceptance**: There is resistance from some communities and environmental groups to the storage of CO_2 underground, particularly related to concerns about leaks or other environmental risks.

- **Recent Developments**:
  - o At COP29, several countries may discuss increased investment in large-scale CCS infrastructure, as well as support for research and development to reduce costs and improve efficiency.
  - o The EU has been promoting CCS as part of its "Green Deal" and may call for greater collaboration in establishing cross-border pipelines and storage networks to facilitate wider adoption.

2. Carbon Capture, Utilization, and Storage (CCUS)

Carbon Capture, Utilization, and Storage (CCUS) goes a step further by not only storing the captured CO_2 but also finding ways to use it as a resource. CCUS technologies can convert CO_2 into valuable products, such as chemicals, fuels, and building materials, thus providing an economic incentive for carbon capture.

- **Key Uses of Captured CO_2**:
  - o **Enhanced Oil Recovery (EOR)**: CO_2 can be injected into oil reservoirs to increase oil production, though this has been controversial due to concerns over long-term emissions.
  - o **Green Fuels**: CO_2 can be used in the production of synthetic fuels, such as synthetic natural gas, methanol, or even jet fuel, by combining it with renewable hydrogen.

- o **Construction Materials**: Captured CO_2 can be used to produce carbon-negative building materials like carbonated concrete.

- o **Agriculture**: CO_2 is sometimes used in greenhouses to enhance plant growth.

- **Recent Developments**:

  - o The rise of "green hydrogen" projects, which utilize renewable energy to produce hydrogen from water, offers an opportunity for CCUS to produce low-carbon hydrogen. Many countries are focusing on integrating CCUS with hydrogen production for both industrial use and energy storage.

- **Challenges**:

  - o **Technology Readiness**: Many CCUS utilization methods, especially in fuel and chemical production, are still in early-stage development and not yet commercially scalable.

  - o **Economic Viability**: For CCUS to be economically viable, the cost of CO_2 capture must be sufficiently low, and there must be a market for the products made from captured CO_2.

3. Carbon Offsets

Carbon offsetting allows individuals, businesses, and governments to compensate for their emissions by funding projects that reduce or remove CO_2 from the atmosphere elsewhere. These projects can range from reforestation and afforestation to renewable energy projects and methane capture from landfills.

- **Types of Carbon Offset Projects**:

  - o **Nature-Based Solutions**: These involve projects such as forest conservation, reforestation, and soil carbon sequestration. They are often considered cost-effective and provide additional benefits like biodiversity conservation and enhanced water resources.

- o **Renewable Energy**: Investing in wind, solar, and hydropower projects that displace fossil fuel-based electricity generation is another form of carbon offset.

- o **Methane Capture**: Methane from landfills, agricultural waste, or coal mines can be captured and either used as energy or flared to reduce its climate impact.

- o **Energy Efficiency**: Projects that improve energy efficiency in buildings, industries, or transportation can be considered offset projects if they result in GHG reductions.

- **Key Principles**:

  - o **Additionality**: The offset project must result in emissions reductions that would not have occurred otherwise.

  - o **Permanence**: The emission reductions must be durable over the long term. For example, carbon sequestration in forests must be protected from deforestation or degradation.

  - o **Verification and Monitoring**: Projects must be independently verified to ensure that they achieve the claimed reductions in emissions.

- **Carbon Offset Markets**:

  - o **Compliance Markets**: These are created under international agreements, such as the Paris Agreement, and national carbon pricing programs. For example, countries and companies subject to emissions caps can buy carbon offsets to meet their targets.

  - o **Voluntary Markets**: Companies and individuals voluntarily buy offsets to neutralize their emissions.

These markets can support a wide range of offset projects but can be more difficult to regulate and standardize.

- Recent Developments:

  o The global carbon market has expanded significantly in recent years, and at COP29, there may be further discussion on how to enhance transparency, reduce fraud, and improve the quality of offset projects.

  o A key topic will likely be how to integrate carbon offset mechanisms with global carbon pricing frameworks and ensure that offset credits contribute to real, permanent reductions in global emissions.

4. Challenges and Criticisms of Carbon Capture and Offsetting

- **Over-Reliance on Technology**: Critics argue that focusing too heavily on CCS and offsetting could divert attention and resources from more immediate and effective mitigation actions, such as transitioning to renewable energy, reducing fossil fuel consumption, and improving energy efficiency.

- **Environmental Integrity**: There are concerns about the integrity of some carbon offset projects, particularly those involving forest conservation, as they can be difficult to monitor and enforce, especially in developing countries.

- **Equity Issues**: Some argue that carbon offsetting can be used by high-emission countries and companies as a way to "buy" their way out of responsibility, without making the necessary changes to their domestic emissions.

- **Long-Term Risks**: While CCS and offsets are seen as vital tools, their long-term effectiveness is still debated, particularly in terms of leakage (CO_2 escaping from storage sites) and the uncertain permanence of carbon sequestration projects.

Chapter 7

Climate Finance

Key Financing Mechanisms

At COP29, **financing mechanisms** will continue to play a pivotal role in supporting both **mitigation** and **adaptation** efforts for climate change, especially in developing countries that may lack the resources and infrastructure to meet the targets of the Paris Agreement. The **Green Climate Fund (GCF)** and the **Adaptation Fund** are two of the key financing mechanisms that will be discussed in detail, along with other funding sources, during the conference. Here's an overview of the most prominent mechanisms:

1. Green Climate Fund (GCF)

The **Green Climate Fund** was established in 2010 during COP16 in Cancun, Mexico, as a financial mechanism under the UN Framework Convention on Climate Change (UNFCCC). Its mission is to support the **transition to low-emission** and **climate-resilient development** by providing funding to developing countries.

- **Objective**: The GCF aims to deliver **climate finance** for both **mitigation** (emission reduction) and **adaptation** efforts, helping vulnerable nations to cope with climate change impacts.

- **Funding Mechanism**:

  o **Mitigation**: The GCF supports projects that reduce emissions through renewable energy, energy efficiency, and low-carbon infrastructure.

  o **Adaptation**: It also funds projects that enhance resilience to climate impacts, such as building sustainable agriculture practices, improving water management, and strengthening infrastructure.

- **Scale of Funding**: The GCF has committed to raising **$100 billion per year** by 2020 (though this target has been delayed) to support climate efforts in developing countries.

- **Governance**: The GCF operates under the guidance of the **Board of Directors**, and funding is allocated based on **country-driven proposals**. Developing countries submit funding proposals, which are then evaluated for their alignment with climate goals.

- **Recent Initiatives**:

    o The GCF has played a key role in funding **climate adaptation programs** for vulnerable regions such as small island developing states (SIDS) and least developed countries (LDCs).

    o Notable projects include **disaster resilience infrastructure**, **climate-smart agriculture** programs, and **renewable energy projects** that replace fossil fuel-based power sources.

- **Challenges**:

    o Despite its substantial funding, the GCF faces criticism for slow disbursement of funds and challenges in reaching the most vulnerable communities. It is also under pressure to meet the target of mobilizing $100 billion per year in a sustainable way.

2. Adaptation Fund

The **Adaptation Fund** (AF) was established under the Kyoto Protocol in 2001 to finance adaptation projects in developing countries that are particularly vulnerable to the effects of climate change. The AF was later incorporated into the **Paris Agreement** to continue its mission and play a role in global adaptation finance.

- **Objective**: The **Adaptation Fund** is dedicated solely to funding **adaptation projects** that help developing countries cope with the adverse effects of climate change, such as extreme weather events, rising sea levels, and changing agricultural patterns.

- Funding Mechanism:
  - The AF supports initiatives that enhance **climate resilience** through programs focused on water security, sustainable land management, agriculture, biodiversity conservation, and disaster risk management.

- Source of Funding:
  - A significant portion of the **Adaptation Fund's** resources comes from a **2% levy** on the proceeds of **carbon market transactions** under the Kyoto Protocol, in addition to voluntary contributions from governments and private donors.

- **Funding Scale**: While the Adaptation Fund's resources are smaller than those of the GCF, it is seen as a key source of financing for urgent and immediate adaptation needs, particularly in **LDCs** and **SIDS**.

- **Governance**: Like the GCF, the Adaptation Fund is governed by a **Board of Directors**, which oversees its operations and ensures that projects align with the principles of climate justice.

- Recent Initiatives:
  - The Fund has supported projects that build **climate resilience** in sectors like **farming, coastal management, water conservation**, and **disaster preparedness**. For instance, it has funded initiatives to improve **rainwater harvesting, flood protection infrastructure**, and **drought-resistant crop varieties** in vulnerable communities.

- Challenges:
  - The Adaptation Fund is often criticized for the **limited availability of financing** relative to the vast adaptation needs in developing countries. Additionally, its funding is generally insufficient to meet the growing demand for adaptation finance in the face of increasingly severe climate impacts.

3. Global Environment Facility (GEF)

The **Global Environment Facility** (GEF) serves as another important financing mechanism for addressing global environmental challenges, including climate change. It operates as a financial mechanism of the **UNFCCC** and focuses on funding projects that address both **mitigation** and **adaptation**.

- **Objective**: The GEF aims to support **sustainable development** and **global environmental benefits** by providing grants to developing countries for projects focused on biodiversity, climate change, land degradation, and water resources.

- **Funding Mechanism**: The GEF has provided funding for over **4,500 projects** globally, contributing to sustainable energy projects, energy efficiency improvements, and climate resilience efforts, particularly in small island states and LDCs.

- **Challenges**: While the GEF has been instrumental in funding environmental initiatives, it often faces limitations in its budget, which is much smaller compared to other climate finance mechanisms like the GCF.

4. Climate Investment Funds (CIF)

The **Climate Investment Funds** (CIF) are another set of financial mechanisms designed to assist countries in transitioning to low-carbon, climate-resilient economies. They include the **Clean Technology Fund (CTF)** and the **Strategic Climate Fund (SCF)**.

- **Objective**: The CIF provides funding for both **mitigation** and **adaptation** projects, focusing on low-carbon energy technologies, sustainable infrastructure, and climate-resilient development.

- **Funding Mechanism**: The CIF mobilizes public and private finance for large-scale investments and leverages concessional finance to attract private capital to key sectors such as renewable energy, energy efficiency, and urban resilience.

- **Recent Initiatives**: Notable projects supported by the CIF include **solar energy installations** in Africa, **clean energy technologies** in Asia, and **climate-resilient infrastructure** in Latin America.

5. Private Sector and Multilateral Development Banks (MDBs)

While public funding mechanisms like the GCF and Adaptation Fund are crucial, the **private sector** and **multilateral development banks (MDBs)** are increasingly being called upon to help finance the global transition to a low-carbon economy.

- **Private Sector Engagement**: Initiatives such as **green bonds, climate-themed investment funds,** and **corporate sustainability efforts** can attract significant private investment in clean energy projects, sustainable agriculture, and climate adaptation technologies.

- **MDBs' Role**: Multilateral development banks, including the **World Bank** and the **Asian Development Bank (ADB)**, are actively financing climate projects, with increasing efforts to integrate climate resilience into their development agendas. MDBs often work alongside the GCF to channel finance into vulnerable countries and sectors.

6. Loss and Damage Fund

A new financial mechanism, the **Loss and Damage Fund**, is expected to be a major focus at COP29. This fund was established to address the **losses and damages** that vulnerable countries face due to **climate impacts** that cannot be avoided through mitigation or adaptation efforts.

- **Objective**: The Loss and Damage Fund will provide financial support to countries suffering from irreversible climate impacts, including rising sea levels, extreme weather events, and biodiversity loss.

- **Funding**: The fund will rely on both **public and private financing**, with contributions from developed nations that are historically responsible for the bulk of global emissions.

Role of public vs. Private funding in climate action

The role of **public** versus **private funding** in climate action is a critical issue in the global response to climate change. Both sectors play complementary roles in financing **mitigation** and **adaptation** efforts, but they come with distinct advantages, challenges, and opportunities. The success of global climate goals hinges on the **collaboration between public and private sector funding** to mobilize sufficient resources to tackle the complex and multifaceted nature of climate change.

1. Public Funding in Climate Action

Public funding refers to financial resources provided by governments and international organizations to support climate-related projects, often through grants, loans, and investments. These funds are typically allocated through **international climate financing mechanisms, development banks, bilateral aid**, and national budgets.

Advantages of Public Funding:

- **Climate Justice and Equity**: Public funding is often used to support the **most vulnerable countries** and communities, especially those with limited capacity to finance climate action themselves. It can ensure that **climate action is equitable** and targets those who need it the most, such as **small island developing states (SIDS)** and **least developed countries (LDCs)**.

- **Long-Term Investments**: Governments can fund large-scale infrastructure projects that require long-term investment horizons, such as renewable energy grids, **resilient infrastructure**, and **climate-proofing agriculture**.

- **Risk Mitigation**: Public funds can assume more **risk** in high-cost, innovative, or untested areas, especially when private investors are hesitant to fund new technologies or solutions. This could include **carbon capture, climate-resilient agriculture**, or **sustainable infrastructure**.

- **Policy and Regulatory Support**: Governments can create enabling environments for climate action through **policy** and regulatory frameworks, thus fostering a supportive ecosystem for private investments.

- **Adaptation Focus**: Public finance is often directed toward **adaptation** efforts that are not always financially attractive to the private sector, such as **disaster risk management, water security**, and **climate-resilient infrastructure** in vulnerable regions.

Challenges of Public Funding:

- **Limited Resources**: Governments may have limited funds available, especially in the face of competing priorities like healthcare, education, and social welfare. While public funding is essential, it may not be enough on its own to meet the growing demand for climate financing.

- **Bureaucracy and Delays**: Public funding is often subject to **bureaucratic processes**, which can delay the disbursement of funds and hinder timely climate action.

- **Dependence on Political Will**: The availability of public funds can be influenced by political changes, economic crises, and government priorities. This makes long-term funding plans vulnerable to shifts in national policies.

2. Private Funding in Climate Action

Private funding comes from individuals, companies, and financial institutions, including **corporations, private equity, banks, insurance companies**, and **philanthropic organizations**. This funding is crucial for driving the massive investments needed to transition to a **low-carbon** and **climate-resilient economy**.

Advantages of Private Funding:

- **Capital Mobilization**: Private sector investment is key to mobilizing the **huge capital** required to scale up climate solutions, especially in sectors like **renewable energy, energy efficiency**, and **sustainable infrastructure**. For instance, private capital is essential for funding the **global transition to solar, wind, and other clean energy technologies**.

- **Innovation and Efficiency**: Private companies often drive **innovation** in clean technologies and business models. Their involvement can speed up the development and scaling of new technologies, such as **electric vehicles, battery storage**, and **smart grids**, and bring **efficiency** to climate action projects.

- **Risk Appetite**: Private investors may be more willing to take risks in developing and emerging markets where climate impacts are most pronounced. They can invest in high-risk, high-reward projects, such as new **carbon capture technologies** or **climate-smart agriculture innovations**.

- **Co-Financing and Leveraging**: Private funding can help **leverage** public finance by attracting co-financing for large-scale projects. Governments can use their funds to de-risk investments and encourage private sector participation, thus maximizing the impact of every dollar spent.

- **Sustainable Business Models**: Increasingly, companies are integrating **sustainability** into their business models, recognizing the long-term benefits of mitigating environmental risks and investing in climate resilience. Many multinational companies have made **net-zero** pledges, driving demand for **climate-conscious investments**.

Challenges of Private Funding:

- **Profit Motive**: Private investments often prioritize **short-term returns**, which may not always align with the long-term goals of climate action, especially for projects requiring significant upfront investment and a slow return, such as **adaptation projects**.

- **Barriers to Investment**: Private investors may be hesitant to invest in certain markets due to **political instability**, **regulatory uncertainty**, and **financial risks** associated with climate change. The lack of clear and consistent climate policies can deter private funding from flowing into climate projects.

- **Focus on Mitigation**: The private sector tends to focus more on **mitigation** efforts, such as renewable energy and energy efficiency, because these sectors are often seen as more financially viable than **adaptation** solutions, which may be harder to monetize.

Public-Private Partnerships (PPPs)

Given the advantages and challenges associated with both public and private funding, **Public-Private Partnerships (PPPs)** are increasingly being seen as a crucial model for advancing climate action. These partnerships combine the strengths of both sectors, ensuring that the **scale, efficiency,** and **social benefits** of public investments are maximized while also tapping into the **capital, innovation,** and **risk-taking abilities** of the private sector.

Key Features of PPPs in Climate Action:

- **Shared Risk and Reward**: In a PPP, both public and private sectors share the financial risks and rewards of climate projects. For instance, a government might fund a portion of a renewable energy project (such as infrastructure) while private investors contribute capital for the technology and operational costs.

- **Leveraging Private Sector Efficiency**: Governments can partner with the private sector to ensure that projects are completed more efficiently and with better innovation, particularly in large infrastructure projects like **sustainable transport, clean energy,** and **resilient cities**.

- **Co-Financing Mechanisms**: Many international funding sources (like the **Green Climate Fund**) encourage co-financing arrangements between public and private entities. This helps ensure that private investors are willing to participate in large-scale projects that contribute to climate goals.

Discussions On Loss and Damage Funding for Affected Nations

Loss and damage funding for climate-affected nations is one of the most pressing issues in international climate negotiations, particularly at the COP (Conference of the Parties) summits. This topic is especially significant for **vulnerable countries** that are already experiencing the severe effects of climate change, such as rising sea levels, more intense storms, droughts, and other extreme weather events. While **mitigation** and **adaptation** are central to global climate action, **loss and damage** focuses on the financial support for countries that face unavoidable impacts of climate change, even after efforts to reduce emissions and build resilience.

1. Understanding Loss and Damage

Loss and damage refer to the **irreparable losses** and **impacts** experienced by countries due to climate change, beyond what can be adapted to. It encompasses both:

- **Losses**: The destruction of natural systems (e.g., biodiversity loss), cultural heritage, and economic assets, such as the loss of agricultural productivity or livelihoods.

- **Damage**: The **physical damage** to infrastructure, homes, and property caused by climate-related events like floods, storms, and wildfires.

The loss and damage agenda is distinct from mitigation and adaptation in that it addresses the impacts of climate change that are **already happening** and are **beyond the capacity** of communities to adapt to or prevent.

2. Key Elements of Loss and Damage Funding

a. The Mechanisms of Loss and Damage Finance

The **loss and damage financing** mechanism aims to provide **financial assistance** to countries that have suffered due to the irreversible consequences of climate change. Discussions around loss and damage funding have evolved over time, especially since the **Warsaw International Mechanism (WIM)** was established at **COP19 (2013)** to address loss and damage in a structured way.

The **WIM** has three components:

1. **Insurance and Risk Reduction**: Aimed at providing tools to protect vulnerable communities from future climate impacts, such as **climate risk insurance, early warning systems,** and **disaster risk reduction strategies.**

2. **Financial Support**: Direct financial assistance for countries suffering from loss and damage, including the creation of a **Loss and Damage Fund.**

3. **Research and Coordination**: Focuses on better understanding the social, economic, and environmental consequences of climate change and creating **data-driven policies** to address loss and damage.

The **Loss and Damage Fund**—a key focus of the discussions at COP28 and COP29—would provide **direct support** to countries that are particularly vulnerable to the consequences of climate change.

b. Sources of Loss and Damage Funding

Funding for loss and damage can come from a variety of sources:

- **Developed Countries**: High-income, historically industrialized countries are expected to contribute the most to loss and damage funding due to their historical responsibility for the majority of global greenhouse gas emissions.

- **Private Sector Contributions**: Financial institutions, insurance companies, and corporations are increasingly seen as potential contributors to loss and damage financing, particularly through mechanisms like **climate risk insurance** and **green bonds.**

- **Climate Finance Institutions**: Organizations like the **Green Climate Fund (GCF)** and the **Adaptation Fund** may also play a role in providing financial support for loss and damage.

- **International Partnerships**: Bilateral agreements and regional cooperation can also contribute, particularly for countries affected by transboundary issues such as **flooding, wildfires,** or **water scarcity.**

At **COP27**, a **loss and damage fund** was officially established to help countries affected by climate impacts. This fund represents a breakthrough after decades of negotiations and provides a platform for supporting vulnerable countries facing extreme impacts.

c. Innovative Financing Mechanisms

To better address loss and damage, innovative financing mechanisms are being explored. These include:

- **Debt-for-climate swaps**: Allowing vulnerable nations to trade debt for investments in climate adaptation or loss and damage mitigation.

- **Climate insurance**: Expanding and scaling **climate insurance** programs, such as the **Global Index Insurance Facility**, to provide quicker, more reliable payouts to countries after natural disasters.

- **Sovereign insurance**: Countries could purchase insurance to cover climate-related losses, ensuring that funds are available immediately after a disaster.

3. Key Discussions and Debates at COP on Loss and Damage

a. Liability and Justice

A central issue in loss and damage discussions is **climate justice** and **liability**. Many vulnerable countries, particularly **small island developing states (SIDS)** and **least developed countries (LDCs)**, argue that developed nations, which historically contributed the most to global emissions, should bear greater responsibility for the costs associated with loss and damage. There is a long-standing debate about the **moral obligation** of high-emitting nations to financially compensate those who are disproportionately affected by climate change, despite contributing the least to its causes.

b. Predictability and Accessibility of Funds

One of the challenges with loss and damage financing is ensuring that funds are **predictable, adequate**, and **accessible**. A common concern among vulnerable nations is that financing mechanisms should be **flexible**, allowing funds to be quickly disbursed in the aftermath of a disaster, without long bureaucratic delays. Ensuring that the **funds reach the most affected** populations and are used effectively is a major point of contention, as administrative burdens and **reporting requirements** may slow down disbursements.

c. Political Will and Funding Commitments

There is ongoing debate about how to **mobilize sufficient funding** and ensure that high-emitting countries fulfill their commitments. In 2009, at **COP15 in Copenhagen**, developed countries pledged to mobilize **$100 billion per year** by 2020 for climate finance, including adaptation and loss and damage. However, there has been criticism that these funds have often fallen short, especially in terms of **loss and damage**. The gap between pledges and actual funding continues to be a significant issue at COP negotiations.

d. Funding for Loss and Damage vs. Adaptation

While adaptation funding is vital for helping countries reduce vulnerability to climate impacts, loss and damage requires funding for situations where impacts are **irreversible** or beyond adaptation. The challenge is ensuring that **loss and damage financing** doesn't simply become a reallocation of **adaptation funds** but is treated as a separate priority that requires new and additional funding streams.

4. Loss and Damage Funding at COP29: Expectations and Focus

As countries prepare for **COP29**, the expectations for **loss and damage funding** are high, with a focus on:

- **Operationalizing the Loss and Damage Fund**: Ensuring that the **Loss and Damage Fund** is operational, with clear mechanisms for accessing the funds, and setting up **transparent governance structures**.

- **Increasing Contributions**: Governments, particularly from developed nations, are expected to make **greater financial commitments** to ensure that loss and damage funding reaches those who need it most.

- **Expanding Insurance and Risk Reduction Tools**: There will likely be discussions on scaling up **climate risk insurance** and **disaster preparedness programs** to ensure that vulnerable nations have the financial tools they need to recover from climate-related disasters.

The focus will be on **mobilizing funds**, ensuring **timely access**, and ensuring that the **funding targets the most vulnerable** countries and communities. There will also likely be calls for **greater equity** in how these funds are distributed and used.

Chapter 8

Technology and Innovation in Climate Solutions

Role Of Renewable Energy, Carbon Capture, And Green Tech

At **COP29**, the role of **renewable energy**, **carbon capture**, and **green technologies** will be crucial as countries work to meet climate targets and reduce global emissions. These technologies are seen as essential for achieving the **Paris Agreement's goal** of limiting global temperature rise to well below 2°C, and ideally to 1.5°C, compared to pre-industrial levels. The discussions at COP29 will focus on scaling up these solutions to address both **mitigation** and **adaptation** needs, ensuring a **sustainable, low-carbon future** for all.

1. Renewable Energy: Powering the Future

a. Importance of Renewable Energy

Renewable energy is at the forefront of climate action because it provides an alternative to **fossil fuels**, which are the largest source of greenhouse gas emissions globally. Solar, wind, hydroelectric, geothermal, and biomass are among the most promising **renewable energy sources** for the future. The global energy transition is necessary to reduce emissions from the energy sector, which is responsible for approximately **two-thirds of global CO2 emissions**.

At **COP29**, renewable energy will be a major topic, with focus on:

- **Scaling Up Deployment**: Efforts will focus on how to accelerate the transition to renewable energy at the **global** and **local levels**, including in emerging economies and vulnerable nations. This includes overcoming **financial, technological, and infrastructure barriers** to renewable energy adoption.

- **Grid Modernization and Energy Storage**: The ability to store renewable energy effectively and efficiently is critical for its widespread use. Discussions will include advancements in **battery storage** technologies, **smart grids**, and **energy**

management systems that enable a more **reliable and flexible energy grid** that can integrate renewable sources.

- **International Renewable Energy Cooperation**: Countries will discuss strategies to enhance international cooperation on renewable energy, including **shared infrastructure** and **transnational energy grids** that can enable the exchange of clean power across borders.

b. Promoting Renewables in Emerging Economies

For developing countries, **renewable energy** offers a unique opportunity to leapfrog traditional fossil-fuel-based energy systems. By adopting clean energy technologies, these nations can mitigate emissions while achieving economic growth and development. At COP29, stakeholders will likely discuss:

- **Financial support for renewables** in low-income countries, potentially through multilateral funding mechanisms like the **Green Climate Fund (GCF)**.

- **Technology transfer** to ensure that developing countries have access to the latest renewable energy technologies and the necessary training and expertise.

2. Carbon Capture and Storage (CCS): Reducing Emissions from Industry

a. Role of Carbon Capture in Emission Reduction

Carbon Capture and Storage (CCS) is an essential technology for reducing emissions from sectors that are difficult to decarbonize, such as **heavy industry (steel, cement, chemicals)** and **power generation**. CCS involves capturing CO_2 **emissions** from industrial processes and storing them underground or using them in products like concrete. It is often seen as a **bridging solution** for the hardest-to-abate emissions while other technologies, like renewables, scale up.

At COP29, carbon capture will be a key topic of discussion, especially in the context of:

- **Scaling up CCS capacity**: With countries setting **net-zero targets**, there will be discussions about the need to **expand** carbon capture infrastructure worldwide. Scaling up **CCS projects** will be essential for reaching global emissions reduction targets.

- **Financing CCS development**: Large-scale CCS projects require significant investment. Stakeholders will discuss how to fund these projects, including public-private partnerships, carbon pricing, and international financing.

- **CCS in industrial sectors**: There will be discussions on **decarbonizing industry** by employing CCS in heavy manufacturing processes. This will require not only the **capture technology** but also infrastructure to transport and store CO2 safely.

b. Negative Emissions and Direct Air Capture (DAC)

Direct Air Capture (DAC) is a technology that captures CO2 directly from the atmosphere. This **negative emissions** technology is seen as a potential tool to offset emissions that are difficult or impossible to eliminate. At COP29, the following topics will likely be discussed:

- **Scaling up DAC**: There will be debates about how to rapidly scale DAC technologies and how to make them economically viable at large scales.

- **Integration of DAC with renewable energy**: For DAC to be effective, it needs to be powered by clean energy. The discussions will likely focus on integrating DAC systems with renewable energy to ensure that the captured carbon is permanently stored.

3. Green Technologies: Driving Innovation Across Sectors

a. Clean Transportation Technologies

Transportation is a major contributor to global greenhouse gas emissions, particularly in **road transport** (cars, trucks) and **aviation**. At COP29, green technologies in transportation will be a major focus, including:

- **Electric vehicles (EVs)**: The adoption of electric cars, trucks, buses, and trains is essential for decarbonizing transportation. Countries will discuss scaling up **EV infrastructure** (e.g., charging stations) and creating incentives for EV adoption.

- **Hydrogen fuel cells**: Hydrogen is seen as a promising alternative fuel, especially for **heavy-duty vehicles, shipping,** and **aviation**. The potential of **green hydrogen** (produced using renewable energy) will be a significant topic of conversation.

- **Sustainable aviation fuels (SAFs)**: Discussions will focus on accelerating the production of SAFs, which can reduce emissions from the aviation sector.

b. Energy Efficiency and Smart Technologies

Energy efficiency is one of the most cost-effective ways to reduce emissions. COP29 will likely see a focus on:

- **Energy-efficient technologies** for buildings, industrial processes, and appliances.

- The use of **smart technologies**, such as **smart meters** and **building management systems**, to optimize energy use and reduce waste.

c. Carbon-Free Manufacturing and Circular Economy

The transition to **low-carbon manufacturing** and a **circular economy** will be central to green tech discussions. Technologies that promote:

- **Recycling** and **reusing materials**, such as using waste carbon for building materials (carbon-capture-based concrete).

- **Energy-efficient industrial processes** that reduce emissions while boosting productivity.

d. Nature-Based Solutions

Nature-based solutions (NBS) are becoming a key part of the green tech conversation, particularly in **carbon sequestration**. COP29 will discuss how to:

- Leverage **forests, wetlands**, and **soil carbon** to absorb CO2.

- Integrate **sustainable land management** practices into climate action plans.

4. Cross-Cutting Issues in Green Technologies

a. Financing Green Technologies

One of the key barriers to scaling renewable energy, carbon capture, and green technologies is financing. **COP29** will see discussions on:

- How to mobilize finance for green tech innovation, especially in **developing countries**.

- The role of **carbon markets, green bonds**, and **public-private partnerships** in supporting green tech development and scaling.

b. Global Cooperation and Technology Transfer

Another key discussion at COP29 will be how to enhance **technology transfer** from developed to developing countries. This will include:

- Sharing **knowledge, best practices**, and **technical expertise**.

- Facilitating **access to affordable green technologies** for developing nations to accelerate their energy transition and build climate resilience.

Breakthrough Technologies Discussed at COP29

At **COP29**, the focus will be on **breakthrough technologies** that can accelerate climate action and help achieve the goals of the **Paris Agreement**. These technologies have the potential to revolutionize how we generate energy, reduce emissions, and adapt to the impacts of climate change. Here are some of the **key breakthrough technologies** that are likely to be discussed at COP29:

1. Direct Air Capture (DAC) and Negative Emissions Technologies

Direct Air Capture (DAC) is one of the most talked-about **negative emissions technologies** at COP29. These systems pull CO2 directly from the atmosphere and either store it underground or use it in products like synthetic fuels or building materials.

- **Carbon capture from the atmosphere**: DAC technologies, still in the early stages of scaling, could become a major tool for offsetting emissions from sectors where decarbonization is difficult. At COP29, discussions may center on **advancing DAC technologies, reducing costs**, and **scaling up operations**.

- **Integration with renewable energy**: For DAC to be effective, it must be powered by clean energy. COP29 will explore how to combine DAC with **solar** and **wind** energy to create a **carbon-negative energy solution**.

2. Green Hydrogen

Hydrogen, particularly **green hydrogen** (produced using renewable energy), is seen as a key solution for **decarbonizing sectors** that are hard to electrify, such as **heavy industry** (steel, cement, chemicals) and **transportation** (aviation, shipping, trucks). At COP29, discussions may focus on:

- **Scaling up green hydrogen production**: Countries and companies will likely discuss how to ramp up **green hydrogen** production, including how to overcome challenges in supply chains, production costs, and infrastructure.

- **Hydrogen for heavy-duty transport**: Hydrogen-powered **fuel cells** may be explored as an alternative to batteries for **long-range heavy vehicles** like trucks, ships, and airplanes.

- **International hydrogen hubs**: Nations may discuss creating **hydrogen hubs** for large-scale production and export, particularly for **countries** with abundant **renewable energy resources**.

3. Artificial Intelligence (AI) in Climate Action

Artificial intelligence (AI) has the potential to accelerate climate action by optimizing energy use, improving climate models, and enhancing the efficiency of clean technologies. Key discussions at COP29 may include:

- **AI for energy efficiency**: AI can help optimize electricity grids, industrial processes, and **smart buildings** to minimize energy use and reduce emissions.

- **AI in climate modeling**: AI is increasingly used to create **more accurate and reliable climate models**, allowing better forecasting of climate impacts and helping policymakers make informed decisions.

- **AI for monitoring and reporting**: AI can help track and monitor **greenhouse gas emissions**, ensuring more accurate **carbon accounting** and facilitating the **reporting of Nationally Determined Contributions (NDCs)**.

4. Advanced Carbon Capture, Utilization, and Storage (CCUS)

Carbon capture, utilization, and storage (CCUS) technologies are critical to decarbonizing heavy industries and sectors that produce large amounts of CO_2 emissions. At COP29, breakthroughs in CCUS will be discussed, such as:

- **Utilization of captured CO2**: Innovations in using captured carbon to create **synthetic fuels, building materials**, or other valuable products will be a central theme. This could create a **circular carbon economy** where carbon is continuously reused.

- **Scaling up CCUS infrastructure**: COP29 will likely feature talks on how to expand **CCUS infrastructure**, including new pipelines and storage sites, to handle large volumes of captured CO2, particularly in industrial regions.

5. Solid-State Batteries and Advanced Energy Storage

Energy storage technologies are essential for integrating **renewable energy** into the grid and ensuring a stable, reliable energy supply. **Solid-state batteries** and other advanced storage solutions are breakthrough technologies that could significantly improve energy storage efficiency. Key discussions may include:

- **Solid-state batteries**: These batteries promise greater **energy density, safety,** and **longer lifespans** compared to conventional lithium-ion batteries, making them ideal for **electric vehicles** and **grid storage.**

- **Next-generation storage systems**: Technologies such as **flow batteries** and **compressed air storage** will be explored for their potential to store large amounts of renewable energy for **long durations.**

6. Ocean-Based Solutions (Blue Technologies)

Blue technologies, which leverage **ocean ecosystems** to combat climate change, are expected to gain significant attention at COP29. These include:

- **Ocean-based carbon capture**: Technologies that use the **ocean** to absorb CO2, such as **marine algae farming, ocean fertilization,** and **blue carbon** (restoring coastal ecosystems like mangroves and seagrasses), may be discussed as viable strategies for **carbon sequestration.**

- **Offshore renewable energy**: **Floating wind farms, wave energy,** and **tidal energy** are seen as promising solutions for generating **clean energy** from the oceans. These technologies are still in early stages but could become a critical part of the global clean energy transition.

7. Geoengineering and Solar Radiation Management

While controversial, **geoengineering** technologies that aim to reduce the **effects of global warming** by **modifying the climate system** may be a topic at COP29. Solar radiation management (SRM), which seeks to reflect sunlight back into space, could be discussed in terms of:

- **Feasibility and risks**: Experts will likely discuss the potential for **SRM technologies** such as **stratospheric aerosol injection** or **marine cloud brightening**, weighing the **scientific uncertainties** and **environmental risks** involved.

- **Governance and ethical concerns**: There will likely be debates on the **governance** of geoengineering technologies, including whether they should be used and who would be responsible for their deployment.

8. Blockchain and Climate Action

Blockchain technology, which provides secure and transparent record-keeping, could be used to improve climate action and tracking emissions reductions. At COP29, discussions may revolve around:

- **Blockchain for carbon markets**: Blockchain can make carbon markets more **transparent** and **efficient**, allowing **carbon credits** to be tracked and traded more effectively, ensuring that reductions are real and additional.

- **Decentralized energy systems**: Blockchain can enable **peer-to-peer energy trading** and create more **decentralized** energy systems, empowering individuals and communities to generate, store, and sell renewable energy.

9. Sustainable Agriculture and Precision Farming

Technologies that promote **sustainable agriculture** will be another focus of COP29. Key innovations may include:

- **Precision farming**: Technologies such as **drones, sensors**, and **AI** will be discussed for their ability to optimize water and fertilizer use, reduce emissions, and increase agricultural productivity.

- **Vertical farming and controlled-environment agriculture**: Urban agriculture solutions, such as **vertical farming**, which use less land and water, could become important for feeding the growing global population while reducing the environmental impact of food production.

10. Climate-Resilient Infrastructure and Smart Cities

Smart cities and **climate-resilient infrastructure** will be central to climate adaptation efforts at COP29. Breakthrough technologies include:

- **Building materials**: Innovations in **low-carbon construction materials**, such as **carbon-negative concrete** and **recycled materials**, will be discussed to help reduce emissions in the **construction sector**.

- **Smart grids and sensors**: Technologies that optimize the distribution and consumption of energy in cities, such as **smart grids, demand-response systems**, and **climate-resilient urban planning**, will be key topics in adapting cities to **climate change**.

Global Partnerships in Technology Sharing

At **COP29**, global partnerships in technology sharing will play a pivotal role in accelerating climate action and achieving the goals set out in the **Paris Agreement**. These partnerships foster the exchange of **knowledge**, **resources**, and **innovative solutions** between countries, organizations, and industries to combat climate change effectively. Here's a detailed look at the role and importance of **global partnerships** in **technology sharing** at COP29:

1. The Role of Technology in Global Climate Action

Technology is central to addressing the climate crisis. However, many of the most promising climate solutions—such as **renewable energy technologies**, **energy efficiency measures**, **carbon capture**, and **climate-resilient infrastructure**—require significant **investment**, **expertise**, and **collaboration**. Global partnerships in technology sharing can provide countries, especially those with fewer resources, access to cutting-edge technologies that help mitigate and adapt to climate change.

At COP29, discussions will focus on how to enhance collaboration between **developed** and **developing nations** to ensure that climate technologies are **accessible**, **affordable**, and **scalable** for all.

2. Key Areas of Technology Sharing

A. Renewable Energy Technologies

The rapid transition to **renewable energy** is crucial for decarbonizing the global economy. Some key areas of technology sharing include:

- **Solar and Wind Power**: Technologies related to **solar panel** production, **wind turbine** efficiency, and **battery storage systems** are among the most widely shared technologies. Partnerships can help scale up the manufacturing and installation of these systems, especially in countries that lack the infrastructure or expertise to do so.

- **Green Hydrogen**: The sharing of **green hydrogen** technologies, including production and storage methods, is vital for countries to use hydrogen as a fuel source in **heavy industries, transportation,** and **power generation.**

B. Carbon Capture, Utilization, and Storage (CCUS)

Carbon capture technologies will be essential in mitigating emissions from hard-to-abate sectors. Partnerships for CCUS technology sharing can enable:

- **Joint development** of advanced **carbon capture** solutions.

- **Technology transfer** to countries that need infrastructure to store captured CO2 or use it in **industrial applications** like synthetic fuels, chemicals, and building materials.

C. Sustainable Agriculture Technologies

Climate-smart agricultural practices are crucial for reducing emissions and improving food security. Key technologies include:

- **Precision farming** tools, such as **drones, sensors,** and **AI-driven decision systems,** can help improve water and nutrient efficiency.

- **Sustainable irrigation technologies** and **crop management systems** help farmers adapt to changing weather patterns and conserve resources.

D. Energy Efficiency and Building Technologies

Building energy-efficient **homes, factories,** and **cities** is key to reducing emissions in the **building sector.** Partnerships can focus on:

- **Energy-efficient building materials** and **smart grids** for better energy management in cities.

- **Cooling technologies** that reduce the need for energy-intensive air conditioning in hot climates.

E. Climate Resilient Infrastructure

Technologies related to creating **climate-resilient** infrastructure—such as **flood defenses, water management systems,** and **climate-proof buildings**—will be crucial, particularly in regions that are vulnerable to the impacts of climate change.

- **Green and blue infrastructure**: Sharing techniques for **urban greening, coastal defenses**, and other forms of climate-resilient infrastructure will be vital for protecting vulnerable communities.

3. Multilateral Platforms for Technology Sharing

There are various platforms and initiatives dedicated to fostering international collaboration on technology sharing:

A. The Technology Mechanism of the UNFCCC

The **UNFCCC Technology Mechanism** is designed to facilitate the development and transfer of climate technologies. It consists of:

- The **Technology Executive Committee (TEC)**: Which provides policy guidance and recommendations on the technology needs of developing countries.

- The **Climate Technology Centre and Network (CTCN)**: A network that provides technical assistance to developing countries to facilitate the transfer and deployment of climate technologies.

At COP29, the **CTCN** and **TEC** will likely play a central role in fostering **technology sharing partnerships**, especially between **developed** and **developing countries**.

B. The Green Climate Fund (GCF)

The **Green Climate Fund** is one of the key financing mechanisms for developing countries, supporting projects related to **mitigation** and **adaptation**. At COP29, the **GCF** will continue to play a significant role in:

- **Financing technology transfer**: The GCF will invest in both the **development** and **implementation** of climate technologies in developing nations, facilitating their access to the latest solutions.

C. Mission Innovation

Mission Innovation is a global initiative launched at COP21 to accelerate public and private clean energy innovation. This partnership includes **19 countries** and the **European Union**, which have committed to doubling their clean energy R&D investments. It aims to:

- **Scale up clean energy technologies** and make them affordable and accessible for all nations.

- Share insights on **policy, financing,** and **implementation** to ensure that new clean technologies are developed and adopted at scale.

D. Global Climate Innovation Alliances

At COP29, **alliances of innovation-focused organizations** may discuss efforts to accelerate technological solutions to climate change. These may include:

- **International collaborations** between universities, research institutions, and private companies to share **R&D** and create joint initiatives for new climate technologies.

- **Private sector partnerships** between **technology companies, startups,** and **multinationals** to develop and scale **low-carbon technologies.**

4. Technology Transfer Challenges and Solutions

While global partnerships in technology sharing have great potential, several **challenges** exist, such as:

- **Intellectual Property (IP):** Some countries may have concerns about **protecting intellectual property rights,** which could slow down the transfer of critical technologies. Solutions may include **open-source licensing** and **sharing agreements** that balance **innovation protection** with global access.

- **Financing:** Many developing countries lack the necessary financial resources to access and implement expensive technologies. **Blended finance** models, combining public and private investments, will be discussed to address this issue.

- **Capacity Building:** In many developing nations, there is a lack of **technical expertise** and **infrastructure** to effectively deploy advanced technologies. Partnerships will need to focus on **capacity-building** initiatives such as **training programs, knowledge transfer,** and **technical support** to ensure successful implementation.

5. Public-Private Partnerships

Public-private partnerships (PPPs) will be crucial in the development and scaling of climate technologies. These collaborations will allow **governments** to leverage **private sector innovation** and **finance** while ensuring that **climate goals** are met. These partnerships may include:

- **Governments** providing incentives, subsidies, and regulatory support.

- **Private companies** offering **innovative solutions** and **financial resources** to develop and implement new technologies.

At COP29, governments and businesses may announce new **collaborations** to share technologies for addressing key climate challenges like **energy production, transportation, agriculture,** and **water management.**

6. South-South Cooperation

In addition to **North-South** technology transfer, **South-South cooperation**—which involves collaboration between developing countries—will also play an important role. **Emerging economies** like **China, India,** and **Brazil** have increasingly become sources of technology and expertise for other developing nations. At COP29, discussions may focus on:

- **Technology sharing** between countries in the **Global South,** such as sharing **solar power solutions** between African countries or **climate-resilient farming techniques** in Asia.

- **Regional partnerships** that facilitate the **exchange of knowledge** and **resources** to tackle common climate challenges.

Chapter 9

Nature-Based Solutions and Conservation

Focus On Biodiversity and Ecosystem Conservation

At **COP29, biodiversity** and **ecosystem conservation** will be at the forefront of discussions as part of the broader conversation on climate change and sustainable development. The **interconnectedness** between climate change, biodiversity loss, and ecosystem degradation requires urgent attention and action from all stakeholders. Here's a detailed exploration of the **focus on biodiversity and ecosystem conservation** at COP29:

1. The Link Between Climate Change and Biodiversity Loss

Climate change is one of the greatest threats to **biodiversity,** driving species loss, altering ecosystems, and disrupting ecological balance. At COP29, key issues related to biodiversity will include:

- **Climate change impacts on ecosystems**: Rising temperatures, altered precipitation patterns, and more frequent extreme weather events are disrupting habitats and biodiversity, threatening species' survival.

- **Ecosystem services**: Biodiversity loss undermines essential ecosystem services, such as clean air, water, food, and disease regulation, which are critical to human survival and well-being.

- **Nature-based solutions**: The potential for **nature-based solutions** (NbS) to both mitigate climate change and conserve biodiversity will be a central theme. **Protecting** and **restoring ecosystems,** such as **forests, wetlands, mangroves,** and **coral reefs,** can sequester carbon, enhance resilience, and protect biodiversity.

2. The Role of Biodiversity in Climate Change Mitigation and Adaptation

Biodiversity plays a dual role in both **mitigation** and **adaptation** to climate change:

A. Mitigation through Ecosystem Conservation

- **Carbon sequestration**: Ecosystems like **forests, wetlands**, and **mangroves** act as **carbon sinks**, absorbing large amounts of carbon dioxide (CO_2) from the atmosphere. **Deforestation**, land degradation, and the destruction of these ecosystems reduce their ability to sequester carbon, exacerbating climate change.

- **Reducing emissions from land use**: Protecting and restoring ecosystems, including through **sustainable agriculture** and **forest management practices**, can help reduce **emissions from land-use change**.

B. Adaptation through Biodiversity Protection

- **Enhancing resilience**: Biodiverse ecosystems are better able to withstand and recover from climate impacts such as storms, floods, and droughts. For example, **mangrove forests** act as natural buffers against storm surges and rising sea levels, while **coral reefs** protect coastal areas.

- **Livelihoods and food security**: Ecosystem conservation can support local livelihoods by maintaining **sustainable fisheries, agriculture**, and **forest-based resources** that are crucial for communities around the world.

3. The Post-2020 Global Biodiversity Framework (GBF)

At COP29, the progress and implementation of the **Post-2020 Global Biodiversity Framework** will be a key focus. This framework, adopted under the **Convention on Biological Diversity (CBD)** at **COP15** in December 2022, sets the global agenda for biodiversity conservation through 2030. The key targets of the GBF include:

- **30x30**: One of the most ambitious targets is the conservation of **30% of the planet's land and oceans** by 2030, to halt biodiversity loss and protect vital ecosystems.

- **Resource mobilization**: Increased financial resources for biodiversity conservation, including a commitment to raise $200 billion per year for biodiversity from both public and private sources.

- **Mainstreaming biodiversity**: The need for biodiversity to be integrated into **sectoral policies** (agriculture, forestry, fisheries, etc.) and national development strategies.

Discussions at COP29 will focus on the **progress** of countries toward these targets and how to **accelerate implementation** of the GBF, particularly in the face of climate change.

4. The Role of Indigenous Knowledge and Community-Led Conservation

Indigenous communities around the world have long been stewards of **biodiversity** and **ecosystems**. At COP29, the recognition of **Indigenous Knowledge Systems** (IKS) and **community-led conservation** will be crucial in discussions on **biodiversity conservation**. Key points will include:

- **Respecting Indigenous rights**: Ensuring that **Indigenous peoples** are involved in decision-making processes around land-use policies, conservation efforts, and biodiversity protection. This includes respecting their land rights and giving them a voice in policy discussions.

- **Community-driven conservation**: Indigenous communities are often the best protectors of ecosystems due to their deep understanding of local environments. Their role in **ecosystem restoration** and **protected areas** will be emphasized at COP29.

5. Integrating Biodiversity into Climate Finance

Biodiversity conservation is not only about protecting ecosystems but also requires substantial **financial investments**. At COP29, discussions will center around:

- **Financing for biodiversity**: Increasing **funding for biodiversity** through mechanisms such as the **Green Climate Fund** (GCF), **Global Environment Facility** (GEF), and other innovative funding models.

- **Biodiversity-inclusive financing**: Ensuring that financial mechanisms for **climate action** also support **biodiversity goals**. This includes integrating biodiversity conservation into broader **climate adaptation** and **mitigation projects**.

- **Public-private partnerships**: Engaging the private sector in funding biodiversity projects, including through **nature-based solutions** and **sustainable business practices** that contribute to ecosystem protection.

6. The Role of Protected Areas and Restoration Initiatives

Protected areas and **ecosystem restoration** will be key topics at COP29. These strategies are central to halting biodiversity loss, mitigating climate change, and improving resilience to climate impacts.

- **Increasing protected areas**: Expanding the network of **protected areas**—such as national parks, marine protected areas, and forest reserves—is vital for safeguarding ecosystems and species.

- **Restoration of degraded ecosystems**: Large-scale **ecosystem restoration** efforts, such as the **UN Decade on Ecosystem Restoration (2021–2030),** will be highlighted as solutions to both **climate change** and **biodiversity loss**. Restoration of degraded land, forests, and marine ecosystems can enhance carbon sequestration and support biodiversity recovery.

7. Nature-Based Solutions (NbS) for Climate and Biodiversity

Nature-based solutions (NbS), which harness the power of nature to address climate change, are increasingly being recognized as crucial for both climate mitigation and adaptation. At COP29, NbS will be discussed in the following areas:

- **Forest conservation and restoration**: Forests are key players in both **carbon sequestration** and **biodiversity conservation**. Programs focused on **reducing deforestation, reforesting,** and **conserving forests** will be explored.

- **Coastal and marine ecosystems**: **Mangroves, seagrasses**, and **coral reefs** play critical roles in protecting coastal ecosystems and biodiversity while mitigating **sea level rise** and **storm surges**.

- **Sustainable agriculture**: Practices such as **agroforestry, organic farming,** and **conservation agriculture** will be highlighted as ways to increase food production while maintaining biodiversity.

8. Addressing Biodiversity Loss through the Private Sector

The private sector has a significant role to play in both **reducing biodiversity loss** and **investing in conservation efforts**. Discussions at COP29 will focus on:

- **Corporate responsibility**: Encouraging businesses to **assess** and **reduce** their impact on biodiversity through **sustainable sourcing, supply chain transparency**, and **biodiversity-friendly practices**.

- **Biodiversity and business models**: Encouraging the integration of biodiversity conservation into business models and strategies, and exploring financial incentives for companies that adopt **nature-positive** practices.

9. International Cooperation and Multilateral Efforts

Global cooperation will be essential for addressing biodiversity loss and ecosystem degradation. At COP29, discussions will focus on:

- **International agreements and initiatives**: Strengthening **global treaties** and collaborations on biodiversity, such as the **Convention on Biological Diversity (CBD)**, the **UN Framework Convention on Climate Change (UNFCCC)**, and the **UN Decade on Ecosystem Restoration**.

- **Cross-border efforts**: Addressing transboundary issues related to ecosystem conservation, such as **cross-border wildlife corridors** and **marine protected areas** that span multiple countries.

Forest Conservation, Reforestation, And Marine Protected Areas

At **COP29**, the topics of **forest conservation, reforestation**, and **marine protected areas** (MPAs) are expected to take center stage as part of the broader discussions on **biodiversity** conservation and **climate change mitigation**. These areas of focus are critical for addressing both the immediate threats of **climate change** and **biodiversity loss**, while also contributing to broader **sustainable development** goals. Here's a detailed exploration of each of these topics:

1. Forest Conservation

Forests play a crucial role in both **mitigating climate change** and **preserving biodiversity**. At COP29, the global community is likely to discuss the **importance of forest conservation** as an essential tool in achieving the goals of the **Paris Agreement** and the **Post-2020 Global Biodiversity Framework**.

A. Role of Forests in Climate Mitigation

- **Carbon Sequestration**: Forests are vital **carbon sinks** that absorb a significant portion of **global carbon dioxide (CO2)** emissions. Forests currently store around **30% of the world's carbon** in vegetation and soils. Deforestation and forest degradation release stored carbon, exacerbating global warming.

- **Preventing Emissions**: Halting deforestation and promoting sustainable forest management is crucial to prevent further emissions. Forest conservation can prevent the loss of carbon sequestered in the trees and soil.

B. Conservation Approaches

- **Sustainable forest management**: The focus will be on implementing and scaling up approaches that allow forests to be used for timber and other resources without causing **long-term damage** to their ecosystems. This includes techniques like **selective logging, agroforestry**, and **community-based forest management**.

- **Forest protection**: A key priority will be ensuring the protection of **primary forests**—those that have not been significantly altered by human activities—as they are irreplaceable carbon sinks and biodiversity hotspots.

C. Forest Conservation Mechanisms

- **REDD+ (Reducing Emissions from Deforestation and Forest Degradation)**: The **REDD+ program** is a major mechanism supported by the UNFCCC and aimed at incentivizing forest conservation and restoration in developing countries. At COP29, countries will discuss how to strengthen REDD+ funding and implementation to halt deforestation and promote sustainable land-use practices.

- **Forest-based Carbon Markets**: As part of the broader carbon trading framework, some countries may focus on developing and enhancing **carbon markets** that allow for forest carbon credits, which can be traded to offset emissions from other sectors.

2. Reforestation

Reforestation, the process of **replanting trees** in areas where forests have been lost or degraded, is seen as one of the most effective ways to address **climate change** and **biodiversity loss**.

A. Importance of Reforestation

- **Restoring Carbon Sequestration Capacity**: Reforestation can help restore the capacity of ecosystems to absorb carbon. Planting trees in deforested areas can increase the **carbon sink potential** of landscapes.

- **Biodiversity Recovery**: Reforestation can help restore habitats for wildlife, bringing back biodiversity to areas that have been cleared for agriculture or other human uses. It can also support the recovery of critical ecosystems like **tropical rainforests**, which are vital for global biodiversity.

B. Reforestation Approaches

- **Large-Scale Reforestation**: COP29 is likely to feature calls for large-scale reforestation programs, particularly in regions that have seen massive deforestation, such as the **Amazon Basin**, the **Congo Basin**, and Southeast Asia.

- **Agroforestry**: Promoting **agroforestry**—the practice of integrating trees into agricultural systems—is an important approach for restoring degraded lands, improving soil quality, and enhancing food security while simultaneously sequestering carbon.

- **Native Species Restoration**: It will be emphasized that reforestation efforts must prioritize the planting of **native tree species** to ensure the recovery of local biodiversity and ecological functions. Non-native species can disrupt local ecosystems and biodiversity.

C. Global Reforestation Initiatives

- **The Bonn Challenge**: The **Bonn Challenge**, which aims to restore **350 million hectares of degraded land** by 2030, is a leading international effort. At COP29, countries will likely discuss how to meet and enhance targets for land restoration.

- **The Great Green Wall Initiative**: The **Great Green Wall** project in Africa aims to restore 100 million hectares of land across the Sahel region by 2030, helping to combat desertification and improve livelihoods. Discussions on this initiative will highlight its importance for both adaptation and mitigation.

3. Marine Protected Areas (MPAs)

Marine ecosystems, particularly **coral reefs, mangrove forests**, and **seagrasses**, play critical roles in both **biodiversity conservation** and **climate change mitigation**. Marine Protected Areas (MPAs) are a vital tool for the protection and restoration of these ecosystems.

A. The Role of MPAs in Biodiversity and Climate Change

- **Preserving Marine Biodiversity**: MPAs protect **marine ecosystems** and **species** from overfishing, pollution, and other human activities. They are essential for preserving the biodiversity of marine life, including vulnerable species like **sea turtles, whales**, and **coral reefs**.

- **Climate Change Resilience**: MPAs enhance the resilience of marine ecosystems to climate change impacts. For example, healthy coral reefs act as buffers against **storm surges**, while **mangroves** and **seagrasses** help protect coastlines from **erosion** and **sea level rise**.

- **Carbon Sequestration**: Marine ecosystems like **blue carbon ecosystems** (e.g., **mangroves, seagrasses**, and **salt marshes**) are capable of sequestering large amounts of carbon. The protection and restoration of these areas are important for reducing atmospheric carbon levels.

B. Expansion of Marine Protected Areas

- **Global Targets**: At COP29, a significant topic of discussion will be how to reach the **30x30 goal**—the target of conserving **30% of the world's oceans** by 2030, a key part of the **Post-2020 Global Biodiversity Framework**.

- **Effective Management**: The focus will be on not just creating MPAs but also ensuring they are **effectively managed**. This includes combating illegal fishing, implementing monitoring systems, and securing **sustainable financing** for the long-term maintenance of MPAs.

- **Marine Spatial Planning**: Marine spatial planning (MSP) is a tool for managing the use of marine resources while minimizing impacts on ecosystems. At COP29, there will likely be discussions on how to scale up MSP to ensure that MPAs are strategically located in areas that maximize both biodiversity protection and ecosystem services.

C. Integrating MPAs with Sustainable Fisheries Management

- **Sustainable Fisheries**: The need to balance the expansion of MPAs with **sustainable fisheries management** will be a key discussion. MPAs can provide **refugia** for fish populations, which can spill over into surrounding areas, benefiting local fisheries.

- **Community Involvement**: Effective MPA management requires the **engagement of local communities** in sustainable fishing practices and conservation efforts. Indigenous and local knowledge should be integrated into MPA planning and management.

Indigenous Knowledge and Practices in Climate Resilience

Indigenous knowledge and practices play a crucial role in building **climate resilience** and contributing to **sustainable climate action**. These knowledge systems, which have been developed over centuries and are deeply rooted in the specific cultural, ecological, and spiritual contexts of Indigenous communities, offer invaluable insights for addressing the challenges posed by climate change. At **COP29,** Indigenous perspectives are likely to be highlighted as key components of **climate adaptation strategies** and **biodiversity conservation efforts**.

1. Understanding Indigenous Knowledge and Practices

Indigenous knowledge refers to the **traditional ecological knowledge (TEK)** that Indigenous peoples have developed through their intimate relationship with the environment. This knowledge is often passed down through generations and is closely tied to the **sustainable management of natural resources, land use**, and **biodiversity conservation**.

A. Holistic Worldview

- Indigenous knowledge often adopts a **holistic worldview**, recognizing the interconnectedness of all life forms and ecosystems. This perspective leads to a deep respect for nature, where humans are seen as an integral part of the broader ecosystem.

- Indigenous communities prioritize **ecosystem health** and the **long-term sustainability** of natural resources over short-term gains, ensuring that environmental practices promote the **well-being of both the land and future generations**.

B. Adaptive Capacity

- Indigenous peoples have adapted to changing environments for thousands of years. Their ability to respond to environmental shifts has led to the development of innovative solutions for managing natural resources, even in the face of **climate variability**.

- Indigenous practices, such as the **rotation of agricultural plots**, **water management systems**, and **fire management techniques**, are based on centuries of observations and experimentation with local ecosystems.

2. Indigenous Contributions to Climate Resilience

At COP29, the recognition of Indigenous practices as effective **climate resilience strategies** is expected to be a central theme. Indigenous communities are often among the most vulnerable to climate change, yet they also offer **innovative solutions** for adaptation and mitigation.

A. Agroecology and Sustainable Agriculture

- **Agroecological practices** are a key component of Indigenous climate resilience strategies. These include traditional methods of **crop rotation**, **polyculture** (growing multiple crops in the same space), and the use of **local seeds** that are more resilient to climate extremes.

- Indigenous knowledge of **soil health**, **water conservation**, and **agricultural biodiversity** promotes agricultural systems that are better able to withstand droughts, floods, and other climate-related stresses.

B. Water and Forest Management

- **Water management** is another area where Indigenous knowledge has been crucial. Indigenous peoples have developed sophisticated systems for managing water, such as **rainwater harvesting**, **terracing**, and **irrigation systems** tailored to local climates and ecosystems.

- In **forest management**, Indigenous communities use traditional fire management techniques, including **controlled burns**, to reduce the risk of larger, more destructive wildfires. These practices maintain **biodiversity**, enhance soil fertility, and promote forest regeneration.

C. Marine Conservation and Fisheries Management

- Indigenous peoples who depend on marine resources, such as **coastal communities** and those living on **islands**, have developed sustainable practices for managing marine ecosystems. These include **seasonal fishing practices, marine conservation areas,** and the use of **traditional knowledge** to monitor fish populations and health.

- For example, in the **Pacific Islands,** Indigenous peoples have developed **reef management strategies** that protect coral reefs and other marine ecosystems, crucial for local biodiversity and resilience to **sea level rise** and **ocean acidification.**

3. Climate Change Adaptation and Indigenous Practices at COP29

At COP29, **Indigenous knowledge** is likely to be highlighted as a vital tool for **climate adaptation.** Some key areas where Indigenous practices could be integrated into climate policies include:

A. Recognition of Rights and Land Stewardship

- A central issue is the recognition of **Indigenous land rights.** Indigenous communities often play a critical role in the stewardship of large areas of **forests, coastal zones,** and **biodiversity hotspots.** Ensuring **land tenure rights** and **self-determination** for Indigenous peoples is crucial to maintaining the resilience of ecosystems and addressing climate change.

- Indigenous governance systems, which emphasize **community-led decision-making** and **collective management,** can serve as models for effective climate resilience strategies that respect both **human rights** and **environmental sustainability.**

B. Traditional Early Warning Systems

- Many Indigenous communities have developed **early warning systems** based on the natural environment. For example, in the **Pacific Islands,** the observation of **cloud patterns, wind shifts,** and **bird behavior** has long been used to predict weather changes and natural disasters. These traditional knowledge systems can complement modern scientific approaches to **climate forecasting** and **disaster preparedness.**

C. Cultural Heritage and Resilience

- **Cultural heritage** is often linked to specific landscapes and ecosystems, and Indigenous peoples have a deep **spiritual connection** to the land and water that sustains them. Protecting this cultural heritage is a critical part of **climate resilience**, as it ensures the continuation of Indigenous **livelihoods** and **ways of life** in the face of climate change.

D. Incorporating Indigenous Knowledge into Global Climate Solutions

- At COP29, **collaborative efforts** between **Indigenous communities** and **national governments** are likely to be emphasized. This could include partnerships to scale up **nature-based solutions**, **land restoration**, and **climate adaptation programs** that incorporate both traditional and scientific knowledge.

- Additionally, the role of **Indigenous knowledge** in **climate mitigation**, such as through carbon sequestration in **forests** and **blue carbon ecosystems**, will be explored. These strategies are often less resource-intensive and more **sustainable** than some modern technological solutions.

4. Challenges and Opportunities

While Indigenous knowledge is essential for **climate resilience**, it faces several challenges, including:

A. Threats to Indigenous Rights and Land

- Indigenous communities are often displaced from their ancestral lands due to **climate impacts**, land grabs, and **unsustainable development projects**. These threats undermine their ability to implement climate-resilient practices.

- **Legal frameworks** and international agreements need to ensure that Indigenous rights are respected and that Indigenous communities have a say in decisions about land use and climate action.

B. Recognition of Indigenous Knowledge in Climate Policy

- One of the challenges is integrating **Indigenous knowledge systems** into **mainstream climate policies** and **scientific research**. There is often a gap between traditional ecological knowledge and the dominant scientific discourse, which can lead to the marginalization of Indigenous voices.

- COP29 will likely provide an opportunity to explore **mechanisms** for **systematically incorporating Indigenous knowledge** into **national climate plans, Nationally Determined Contributions (NDCs)**, and **climate financing.**

Chapter 10

Youth and Civil Society Movements

Involvement Of Youth Groups, Activists, And NGOs

At **COP29**, the involvement of **youth groups, activists,** and **non-governmental organizations (NGOs)** is expected to play a significant and increasingly influential role in shaping the direction of climate action. These groups have been at the forefront of pushing for **more ambitious climate policies, social justice,** and **greater urgency** in addressing the climate crisis. Their participation adds **diverse perspectives** and **vital energy** to the negotiations, contributing to a broader, more inclusive approach to global climate solutions.

1. Youth Groups and Activists at COP29

Youth-led movements have become a defining feature of recent climate conferences, driven by the understanding that **young people** will bear the long-term consequences of today's decisions on climate change.

A. Global Youth Movements

- Youth movements such as **Fridays for Future, Youth for Climate,** and **Climate Strike** have gained international prominence, calling for **immediate action** to mitigate climate change and to ensure a **livable future** for younger generations.

- **Greta Thunberg**, the Swedish environmental activist, is one of the most visible figures in these movements, whose **school strike for climate** ignited the global movement of young people demanding urgent action.

- These youth groups are often advocating for **radical shifts** in policy, including pushing for **phasing out fossil fuels, expanding renewable energy,** and **prioritizing climate justice**. At COP29, they are likely to demand **stronger commitments** from **governments** to meet climate goals, particularly the **Paris Agreement** targets.

B. Role of Youth in Climate Advocacy

- Youth groups are likely to continue organizing **demonstrations** and **activist events** during COP29, putting pressure on negotiators and world leaders to take bolder steps toward addressing the climate crisis.

- **Youth delegates** may be involved in formal negotiations or side events, advocating for the inclusion of **youth perspectives** and ensuring that **future generations** are given a seat at the table in decisions affecting the planet's future.

- They often push for **intergenerational equity**, meaning that climate policies should not only address the needs of the current generation but also **protect the rights** and **future well-being** of young people.

C. Youth Engagement and Leadership

- Youth groups at COP29 are likely to focus on themes like **climate education, empowerment**, and ensuring that **youth voices** are heard in both the **negotiating halls** and in the **public discourse** on climate action.

- They are also expected to advocate for **greater inclusion of young people** in **leadership positions** within international climate bodies, as well as in **policy-making processes** at local, national, and international levels.

2. Activists and Climate Justice

Climate activists—both young and from diverse age groups—often champion **climate justice**, focusing on **equity, human rights**, and the disproportionate impacts that climate change has on **marginalized communities** (e.g., **low-income populations, indigenous communities**, and **people in vulnerable regions** such as small island states).

A. Raising Awareness on Climate Injustice

- Activists may highlight the need for **climate policies** that not only mitigate greenhouse gas emissions but also address the **social and economic inequalities** that climate change exacerbates. These can include issues such as **displacement, loss of livelihoods, health impacts**, and **gender inequality**.

- Calls for **climate reparations**—that wealthy nations who have historically contributed the most to global warming should compensate developing nations that bear the brunt of climate impacts—are expected to be a focal point for activist groups.

B. Climate Justice and Equity

- The demand for **climate justice** involves ensuring that the **financial and social burdens** of climate action do not fall disproportionately on the **most vulnerable communities**. This includes prioritizing the **needs of Indigenous peoples, women, youth**, and **low-income nations** in climate policy.

- Activists often call for **loss and damage financing** that compensates countries that suffer the greatest loss due to climate change, and ensure that funding mechanisms are **accessible and fair**.

3. NGOs' Role in COP29

Non-governmental organizations (NGOs) play a vital role in influencing climate negotiations, advocating for policy changes, and mobilizing public support for climate action.

A. NGOs as Advocates for Action

- NGOs, such as **Greenpeace, World Wildlife Fund (WWF), 350.org**, and **Oxfam**, are likely to have an active presence at COP29, advocating for **stronger emissions reductions**, the **expansion of renewable energy**, and the **protection of biodiversity**.

- These organizations often serve as **watchdogs**, scrutinizing governmental positions, and ensuring that countries remain accountable to their **climate commitments**.

- NGOs also contribute to the conference by organizing **side events, workshops**, and **public outreach campaigns** to raise awareness and generate public pressure for **climate action**.

B. Supporting Vulnerable Communities

- Many NGOs focus on the needs of the **world's most vulnerable populations**, such as those living in **low-lying islands, desert regions,** or **deforested areas,** whose livelihoods are particularly susceptible to climate change impacts.

- These NGOs often work to **amplify the voices** of marginalized communities during the negotiations, calling for **fair climate finance, adaptation support,** and **just transitions** to a **green economy.**

C. Partnerships with Governments and Private Sector

- NGOs also facilitate partnerships between **governments, private companies,** and **civil society** to drive innovative solutions in areas like **sustainable agriculture, renewable energy, carbon capture,** and **water management.**

- They often play an important role in **capacity building** and **training,** helping communities implement **climate resilience strategies** and access **adaptation finance.**

4. Collaborative Efforts and Cross-Sector Engagement

At COP29, the collaboration between **youth groups, activists, NGOs, government representatives,** and **private sector actors** will be critical in shaping meaningful outcomes. These groups often **coordinate their efforts** to ensure that **climate action** is not only ambitious but also equitable and inclusive.

A. Coalitions and Alliances

- Many **climate action coalitions** form around shared priorities, such as **the Global Climate Strikes, the We Are Still In coalition,** and **the Climate Action Network,** which bring together a wide range of actors—including NGOs, businesses, and youth groups—to advocate for stronger action at COP negotiations.

- **Cross-sector partnerships** will be crucial in driving **innovative solutions** that integrate **social justice, environmental sustainability,** and **economic resilience.**

Role Of Social Media and Global Awareness Campaigns

The role of **social media** and **global awareness campaigns** at **COP29** is expected to be pivotal in shaping the narrative around climate change, driving public engagement, and holding leaders accountable for their climate commitments. Social media platforms provide an **immediate and powerful channel** for spreading information, mobilizing grassroots movements, and influencing the discourse at the conference. These tools help amplify the voices of activists, organizations, and affected communities, contributing to a **more inclusive and transparent** climate conversation.

1. Social Media's Role in Amplifying Climate Action

Social media platforms like **Twitter, Instagram, TikTok**, and **Facebook** will continue to play an essential role in communicating key messages related to climate action during COP29. They offer a **global stage** for activists, NGOs, and even governments to present their **views, goals, and demands**, and to **raise awareness** about the climate crisis. Here's how social media influences COP29:

A. Real-time Information Sharing

- Social media allows for **instant updates** from the conference, enabling stakeholders and the global public to stay informed about the latest developments, decisions, and negotiations.

- Delegates, media outlets, NGOs, and activists can quickly share **press releases**, **important announcements**, and **highlights** from the event, ensuring that key messages are broadcast globally in real-time.

B. Mobilizing Public Opinion and Activism

- Social media has become a **tool for activism**, with platforms being used to mobilize people for **climate protests, demonstrations**, and **virtual events**. Movements like **Fridays for Future**, which was initially sparked through social media, are likely to have an amplified presence at COP29.

- Hashtags like **#COP29, #ClimateActionNow**, and **#ActOnClimate** are used to **amplify messages** across the globe, helping campaigns trend and reach larger audiences.

These hashtags help create a unified, global voice that advocates for climate justice and action.

- Through **online petitions, calls for action**, and **direct engagement with world leaders** via social media, the public can put pressure on negotiators to adopt **stronger climate policies.**

C. Global Engagement and Education

- Social media serves as a **platform for education**, with influencers, scientists, and environmental organizations sharing **educational content, scientific reports**, and **climate facts** to increase awareness among the general public.

- Infographics, videos, and live-streamed events allow complex climate data and negotiations to be **presented in an easily digestible format**, making climate science and COP discussions more accessible to a global audience.

- Educational campaigns around specific issues, such as **biodiversity loss, climate justice**, or **green energy** solutions, can attract widespread engagement, especially among younger audiences.

2. Global Awareness Campaigns and Advocacy

Global climate campaigns and advocacy efforts have become significantly stronger with the rise of **digital platforms**. Many NGOs, youth organizations, and climate movements use social media to **build global solidarity** and create movements that demand greater climate action. Here's how global awareness campaigns influence COP29:

A. Amplifying the Voices of Marginalized Communities

- Social media can **amplify voices** from communities most affected by climate change, including **small island states, indigenous groups**, and **low-income countries**. By **sharing personal stories** and **on-the-ground experiences**, these communities can bring attention to the **inequities** they face and demand **climate justice.**

- Campaigns focused on **loss and damage**—such as **#LossAndDamage**—raise awareness about the need for **financial support** and **adaptation measures** for vulnerable populations impacted by climate disasters.

B. Cross-Network Collaboration and Action

- Social media helps foster collaboration between **climate justice groups, environmental NGOs, youth activists, scientists**, and **businesses**. These networks can rapidly mobilize to support campaigns, influence policies, and raise funds for **climate resilience projects**.

- At COP29, global campaigns may focus on the **urgent need** for stronger **emission reduction targets, green technology investment, sustainable agriculture**, and **preserving biodiversity**. The collective action sparked by social media campaigns often results in **joint policy advocacy**, as well as the establishment of **climate justice coalitions**.

C. Shaping the Narrative and Public Pressure

- Social media is a tool for **shaping public narratives** around climate change. Activists can quickly challenge misleading information or **lobbying efforts** by vested interests that downplay the importance of addressing the climate crisis.

- In addition to **raising awareness**, online campaigns can influence public pressure, compelling governments and private sector players to **meet their climate pledges** or adopt stronger measures. Calls for **accountability** on social media can have a profound effect on policy-makers, as they are directly in the spotlight.

D. Campaigns Promoting Climate Justice and Equity

- Many campaigns center around the issue of **climate justice**, particularly the unfair impacts climate change has on **developing nations** and **marginalized groups**. Activists push for a **global climate fund** and **loss and damage compensation** to support these communities.

- Campaigns also emphasize the need for **gender-responsive climate policies**, ensuring that **women**—especially in vulnerable regions—have the resources and support needed to address climate challenges.

3. Influence of Climate Influencers and Public Figures

- Public figures, including **celebrities, environmental leaders,** and **scientists,** have used social media to **amplify climate messaging**. Their influence can **reach millions** of followers, encouraging them to take action on climate change, participate in **virtual events,** or advocate for stronger policies.

- Figures like **Greta Thunberg, David Attenborough,** and **Leonardo DiCaprio** are known for using their platforms to call for **urgent action** and to highlight the importance of international agreements such as the **Paris Agreement**.

4. Live-Streaming and Interactive Engagement

- Social media has allowed **live-streaming** of **COP sessions,** making the events more transparent and accessible to people who cannot attend the conference in person. This fosters a more **interactive climate dialogue** between negotiators, activists, and the public.

- **Virtual participation** allows individuals and groups to ask questions, attend digital panels, and interact with stakeholders, further democratizing the conversation and ensuring that those most affected by climate change have a voice in the discussions.

5. Tracking Commitments and Holding Leaders Accountable

- Social media can **track climate pledges** made by governments and corporations, helping to ensure that commitments made during COP29 are **followed through**. Activists and the media use digital tools to **monitor progress** and **publicly report** on any **lack of action** or failure to meet targets.

- The #FollowTheMoney campaign, for instance, is a way for stakeholders to track **financial contributions** to climate adaptation and mitigation efforts, ensuring that the promises made at COP are actually met.

6. Climate Communication and Behavioral Change

- Social media is an essential tool for spreading messages of **sustainable living** and **individual climate action**. Through educational content, social media campaigns encourage **behavioral changes** such as reducing carbon footprints, shifting to renewable energy, reducing consumption, and supporting **green businesses**.

- Influencers and environmental groups can lead **challenges** (e.g., **#PlasticFree**, **#MeatlessMonday**) to encourage sustainable habits that collectively contribute to reducing global emissions.

Grassroots Movements and Community-Led Initiatives

Grassroots movements and community-led initiatives play a critical role in shaping the climate action agenda, particularly during international events like **COP29**. These movements, often led by individuals, local organizations, or activist groups, help ensure that climate change policies are inclusive, just, and responsive to the needs of vulnerable and marginalized communities. They focus on **local-level solutions, empowerment**, and **collaboration**, building momentum for systemic change and influencing global discussions on climate action.

1. Role of Grassroots Movements in Climate Action

Grassroots movements are typically driven by communities most directly impacted by climate change, such as **low-income groups, indigenous peoples**, and **small island nations**. These movements often advocate for **climate justice**, ensuring that the burdens of climate change do not disproportionately affect those who have contributed least to the crisis.

A. Raising Awareness

- Grassroots movements often focus on **educating** local communities about climate change, its impacts, and the actions people can take. They may use **storytelling**, **community meetings**, and **local media** to spread information and build awareness.

- These movements also raise awareness on the **political front**, pressuring governments and international bodies to take action on **mitigation** and **adaptation** policies that consider the needs of the most vulnerable.

B. Advocating for Climate Justice

- Grassroots groups often highlight the **disproportionate impacts** of climate change on marginalized populations. They push for **fair compensation, financial support**, and **climate adaptation resources** for communities that are already experiencing the worst effects, like **flooding, drought**, or **extreme weather events**.

- They campaign for **just transition policies** that move away from fossil fuels in a way that supports workers and communities, particularly in regions reliant on **coal** or **oil industries**.

C. Creating a Global Movement for Action

- Grassroots movements have the power to unite individuals and communities worldwide under common causes. Through **social media, global protests**, and **solidarity campaigns,** they help build **international pressure** on leaders to take bold climate action.

- Movements like **Fridays for Future**, led by **Greta Thunberg**, or **Extinction Rebellion,** have sparked global awareness of the urgency of addressing climate change, demonstrating how grassroots campaigns can bring together people from different countries and backgrounds to call for change.

2. Examples of Successful Grassroots Movements

A. Fridays for Future

- Founded by Swedish teenager **Greta Thunberg, Fridays for Future** is one of the most high-profile climate action movements globally. It began with Greta's **school strike for climate** and has since grown into a **youth-led global movement** advocating for urgent climate action.

- The movement's demands include **stronger climate policies, scientific evidence-based action,** and **global accountability** for emissions reduction. The movement has mobilized millions of students and activists worldwide, organizing **global strikes, protests**, and **advocacy campaigns**.

B. The Sunrise Movement (U.S.)

- The **Sunrise Movement** is a grassroots organization in the United States that advocates for a **Green New Deal**, focusing on both climate change and social justice. The movement has been instrumental in pushing the idea of a **Green New Deal** into mainstream politics and influencing major political figures, particularly among **young Americans**.

- Their campaigns focus on **clean energy jobs, green infrastructure,** and **a just transition for workers** affected by the transition from fossil fuels. They have successfully organized **youth-led protests** and **advocacy events** to bring attention to the climate emergency.

C. 350.org

- Founded by environmentalist **Bill McKibben, 350.org** is a global grassroots movement dedicated to reducing **carbon dioxide levels** in the atmosphere to 350 ppm (parts per million). The movement organizes **climate action campaigns, public demonstrations,** and **education initiatives** to inspire a global transition to renewable energy and **climate justice.**

- 350.org has mobilized millions worldwide, focusing on actions like **divestment from fossil fuels** and **protests against pipeline projects** that contribute to climate change.

D. Indigenous Climate Action

- Indigenous communities, especially those in the **Amazon, Pacific islands,** and **Arctic regions,** have long been at the forefront of **climate resistance.** Groups like **Indigenous Climate Action** in Canada and **The Pacific Islands Climate Change Cooperation (PICCC)** advocate for **land rights, environmental protection,** and **sustainable development** practices.

- These movements emphasize the importance of **indigenous knowledge,** including **traditional ecological practices** that have helped communities live in harmony with nature for centuries. They push for **legal protections** for land and water rights and advocate for **self-determination** in climate policies.

3. Community-Led Adaptation and Mitigation Initiatives

Community-led initiatives are especially important in the **adaptation** and **mitigation** spaces. By drawing on local knowledge, these projects often yield more **effective and sustainable solutions** tailored to the specific needs of a community.

A. Community-Based Renewable Energy Projects

- In regions with **limited access to electricity**, grassroots organizations have worked to implement **community-owned renewable energy** systems, such as **solar panels**, **wind turbines**, and **microgrids**. These projects not only help reduce emissions but also empower communities by providing affordable, **clean energy**.

- For example, **Off-grid Solar Energy Initiatives** in Africa and Asia have provided **renewable energy solutions** to rural villages, improving access to electricity while contributing to **climate change mitigation**.

B. Agroecology and Sustainable Agriculture

- **Agroecology**, a community-led farming approach that combines traditional and scientific knowledge, has gained prominence in grassroots movements as a way to both **adapt to climate change** and **mitigate emissions** from industrial agriculture.

- Programs such as **agroforestry**, **soil regeneration**, and **sustainable water management** help communities **reduce their vulnerability** to droughts, floods, and other climate impacts. They also enhance **biodiversity**, improve soil health, and **sequester carbon**.

C. Community Water Management

- In areas prone to **water scarcity** or **flooding**, community-led water management systems—such as **rainwater harvesting** and **restoration of wetlands**—have proven to be vital for **climate adaptation**. These grassroots projects help communities **manage water resources** sustainably, build **climate resilience**, and reduce the impacts of **climate-induced disasters**.

D. Reforestation and Land Restoration

- **Community-led reforestation** projects are another significant initiative, particularly in regions affected by **deforestation** and **land degradation**. These initiatives not only combat

climate change by absorbing carbon but also restore vital ecosystems.

- In places like **Kenya, Indonesia,** and **Nepal,** communities have led successful projects to plant **native trees,** restore soil quality, and protect **biodiversity.** These efforts are often backed by **local knowledge** and **traditional conservation practices.**

4. Challenges Faced by Grassroots Movements

While grassroots movements and community-led initiatives are vital for climate action, they face several challenges, including:

A. Funding and Resources

- Grassroots organizations often struggle with **limited financial resources,** hindering their ability to scale up initiatives or influence policy decisions effectively. Securing **consistent funding** from governments, foundations, or private donors is crucial for their sustainability.

B. Political and Legal Barriers

- In many regions, political or legal barriers hinder the ability of grassroots groups to push for **environmental reforms.** This can include **restrictive laws, land tenure issues,** or **government opposition** to grassroots environmental activism.

C. Media Attention and Public Support

- Although social media plays a critical role, grassroots movements often face **challenges in breaking through the noise** of mainstream media, where climate change is sometimes marginalized in favor of other topics. Sustaining **public attention** over time is crucial for continuing momentum.

Chapter 11

Challenges and Controversies

Conflicts Over Emissions Targets and Financial Commitments

Conflicts over emissions targets and financial commitments are critical issues that often emerge during international climate negotiations like **COP29**. These conflicts arise from differing national priorities, economic capacities, and political agendas, making consensus-building a challenging task. The complexity of balancing climate action with national economic interests, equity concerns, and financial obligations often leads to contentious debates.

1. Conflicting Emissions Targets: Developed vs. Developing Nations

One of the most significant conflicts in climate negotiations involves the setting of **emissions reduction targets**. These conflicts typically stem from differing responsibilities for past emissions, levels of economic development, and capabilities to address climate change. There are two main dynamics at play:

A. Historical Responsibility of Developed Nations

- Developed nations, including the **United States, European Union**, and **Japan**, have historically contributed the most to **global greenhouse gas emissions** since the **industrial revolution**. These countries are often seen as having a moral obligation to take the lead in reducing emissions.

- Developing nations, including **India, China**, and **African nations**, argue that they should not be expected to make the same level of emissions reductions as developed countries, given their **historical responsibility** for emissions and their current needs for **economic growth** and **poverty alleviation**.

- This has led to disputes about the **fair share** of emissions reductions that each country should be responsible for, especially in the context of **global climate goals** like limiting **global warming to 1.5°C or 2°C**.

B. Differentiated Responsibilities and Capabilities (CBDR)

- The principle of **common but differentiated responsibilities and respective capabilities (CBDR-RC)** has been central to the UNFCCC framework. It recognizes that while all countries should take action against climate change, the level of responsibility and the approach should differ based on their **historical contributions, economic development, and capacities.**

- However, the implementation of this principle remains contentious. Developing countries often argue that developed nations must take more **aggressive action** and provide **financial support** for **climate adaptation** and **mitigation** in the Global South.

C. Fair Share of Emissions Reductions

- At COP29, discussions around **fair share** could be a point of contention, particularly regarding how to allocate emissions reductions across nations. For instance, some emerging economies like **China** and **India**, which are large emitters due to their rapid industrialization, argue that they should not be held to the same standards as developed countries given their **developmental needs.**

- On the other hand, many high-income nations argue that the world's largest emitters must take on stronger commitments, especially as countries like **China** have become the world's largest emitter in recent years.

2. Financial Commitments: Climate Finance for Developing Countries

Another area of major conflict in climate negotiations is the **financial commitments** made by developed countries to assist developing nations in **mitigating** and **adapting** to the impacts of climate change. Developing countries have long argued that financial support is crucial for their ability to take meaningful climate action. However, the gap between what is pledged and what is delivered has been a persistent source of tension.

A. $100 Billion Climate Finance Pledge

- At COP15 (Copenhagen, 2009), developed countries committed to providing **$100 billion per year** in climate finance to developing countries by 2020. This funding was intended to help nations in the Global South adapt to climate impacts, transition to cleaner energy sources, and implement sustainable development goals.

- However, this pledge has faced significant challenges, with developing countries accusing wealthier nations of falling short on both the **amounts promised** and the **transparency** of climate finance disbursements.

- At COP29, it is expected that developing nations will continue to pressure developed countries to meet this funding commitment, and to ensure that it is **new and additional** funding (not diverted from existing aid budgets).

B. Sources and Mechanisms of Funding

- There is ongoing debate over the **sources** of climate finance. While **public funds** from governments are a key source, there is also increasing pressure for **private sector** investments, including **green bonds**, **carbon markets**, and **climate insurance products**.

- However, critics argue that relying on **private finance** risks undermining the fairness and equity of climate finance, as private investors may prioritize returns over the **needs** of vulnerable communities.

- **Loss and damage finance** also remains a key point of contention. Developing countries, particularly small island states and African nations, argue that they should receive financial support for the **irreparable damage** caused by climate change, including **sea-level rise**, **extreme weather events**, and the **destruction of ecosystems**. While progress has been made, including the establishment of a **loss and damage fund** at COP27, the details of its **implementation** and the **amount of funding** remain unresolved.

C. Tracking and Transparency of Funding

- Another point of conflict is the **tracking** and **accountability** of climate finance. Developing nations argue that there needs to be more **transparency** in how climate finance is tracked and reported. They insist on mechanisms to ensure that the financial support promised is **delivered in a timely manner** and that funds are used effectively for climate action.

3. Disagreements Over the Use of Carbon Markets

Carbon markets, including **carbon credits** and **carbon offset mechanisms**, have been a point of conflict in climate negotiations due to their complex nature and questions over their **effectiveness** in achieving real emissions reductions.

A. Carbon Offsets and Market-based Mechanisms

- **Carbon credits** allow companies or countries to offset their emissions by investing in projects that reduce emissions elsewhere, such as **reforestation** or **renewable energy** projects. However, concerns have been raised about the **integrity** of these credits, particularly in terms of whether they represent **real emissions reductions** or merely **paper-based solutions**.

- Some developing countries argue that carbon markets should be an **additional tool**, but not a substitute for **real emissions reductions**. There is also criticism that wealthy nations or corporations could use offsets to avoid reducing their own emissions, instead of investing in genuine decarbonization efforts.

B. Article 6 of the Paris Agreement

- Article 6 of the **Paris Agreement** deals with **market-based approaches**, including **carbon trading** and **international cooperation** to achieve emissions reductions. While these mechanisms can potentially offer flexibility and cost-effectiveness, they also raise questions about **accounting** and **equity**. Developing countries, particularly the **least developed countries (LDCs)**, are wary of these market-based approaches being exploited in ways that might not lead to substantial reductions in emissions or might exacerbate inequality.

4. Geopolitical Tensions and Climate Diplomacy

Global geopolitical tensions can also contribute to conflicts over emissions targets and financial commitments. For example:

A. U.S.-China Rivalry

- The rivalry between **China** and the **U.S.**, the two largest greenhouse gas emitters, can influence the level of ambition at international climate talks. Disagreements over the responsibilities of each country in addressing climate change can stall progress on achieving a global consensus. While both countries have taken steps to reduce emissions, they also have contrasting priorities that can lead to tensions over **target setting** and **financial contributions**.

B. Impact of War and Conflict

- Ongoing geopolitical conflicts, such as the **Russian invasion of Ukraine**, can also undermine climate progress. The energy crisis caused by such conflicts can divert attention and resources away from climate action. Countries may prioritize energy security or national economic interests, slowing down the transition to renewable energy or delaying international climate commitments.

Disparities Between Developed and Developing Nations

The **disparities between developed and developing nations** remain one of the most fundamental and contentious issues in global climate negotiations, including at **COP29**. These disparities arise from differences in **historical emissions, economic development, capacity to mitigate climate change**, and **vulnerability to its impacts**. Understanding these disparities is crucial for addressing the challenges of **global climate action**, as well as ensuring **equity** and **justice** in the implementation of climate solutions.

1. Historical Responsibility for Emissions

A. Developed Nations: Historical Emissions

- **Developed countries**—such as the **United States, European Union, Japan,** and **Canada**—have been responsible for a significant portion of global greenhouse gas (GHG) emissions since the **industrial revolution**. The emissions from these nations have largely contributed to **global warming** and the climate crisis.

- Historically, **industrialization** in developed nations relied heavily on **fossil fuels**, leading to long-term carbon emissions. As a result, these nations have accumulated a large **carbon debt**, which many developing countries argue should be taken into account when setting climate targets.

- This historical responsibility has led to calls from **developing nations** for developed countries to take the lead in **emissions reductions** and provide **financial support** for climate action in the Global South.

B. Developing Nations: Low Historical Emissions, Growing Contribution

- In contrast, many **developing countries**—such as **India, China,** and **Brazil**—have contributed far less to **historical emissions**. However, as these nations rapidly industrialize and urbanize, their **current emissions** are rising.

- **China** is now the **world's largest emitter**, and **India** is also seeing increasing emissions due to its large population and economic development needs. However, these countries argue that their per capita emissions are still low compared to developed countries and that they should not be expected to make the same level of commitments to reducing emissions without **financial and technological support**.

2. Economic Capacity and Development Needs

A. Developed Countries: Greater Resources

- Developed nations generally have **more financial resources, technological capabilities**, and **infrastructural capacity** to take action on climate change. These countries are better positioned to invest in **clean energy technologies**, implement **carbon-reducing policies**, and adapt to climate impacts.

- They also have stronger **financial markets**, more access to **private capital**, and greater **industrial capabilities** to transition to a **low-carbon economy**. As a result, there is an expectation that developed countries should take on more ambitious emissions reduction targets and be held accountable for meeting **financial commitments** to assist developing nations.

B. Developing Countries: Economic Vulnerability

- Developing nations, on the other hand, face significant **economic challenges** and often rely on **fossil fuel-based industries** for growth. Many of these countries are **resource-dependent** economies, with heavy reliance on sectors like **agriculture, mining**, and **energy extraction**.

- They also have pressing needs for **economic development**, including the reduction of **poverty, improved healthcare**, and **energy access**. For many countries in the **Global South**, the **immediate economic needs** often compete with the demands of **climate action**. Hence, they argue that the transition to a **low-carbon economy** should be **just** and **fair**, providing adequate support to avoid hindering their development prospects.

3. Vulnerability to Climate Impacts

A. Developed Countries: Better Adaptation Capacity

- Developed countries typically have more robust infrastructure, disaster-response mechanisms, and financial resources to adapt to the impacts of climate change, such as extreme weather events, sea-level rise, and wildfires.

- While these countries still face significant climate-related challenges (e.g., heatwaves, wildfires, and flooding), their higher levels of wealth and development often provide greater resilience to such impacts.

B. Developing Countries: Highly Vulnerable

- Developing countries, particularly small island states, low-lying coastal nations, and least developed countries (LDCs), are among the most vulnerable to climate change impacts. These countries are often exposed to severe weather events such as hurricanes, flooding, and droughts, which can devastate their economies and communities.

- For many of these countries, climate change represents a threat to survival. For example, Pacific island nations face existential threats from sea-level rise, while African nations are facing increased droughts and food insecurity. As a result, they demand more financial assistance and technological support to help them adapt to these challenges.

- The vulnerability of developing countries is often compounded by poverty, poor infrastructure, and lack of access to resources, which limits their capacity to deal with climate impacts effectively.

4. Climate Finance and Financial Commitments

A. Developed Countries' Financial Responsibilities

- Developed nations have long been expected to provide climate finance to assist developing countries in both mitigation and adaptation efforts. At COP15 (Copenhagen, 2009), developed countries committed to providing $100 billion per year in **climate finance** by 2020 to help developing countries address the impacts of climate change.

- However, this commitment has faced significant challenges, with many developing nations arguing that developed countries have **not met** the promised targets, and that the climate finance provided has often been in the form of **loans** rather than grants, adding to the debt burden of vulnerable nations.

B. Climate Finance Gap and Accountability

- Developing countries continue to push for **more transparency** and **accountability** regarding climate finance. They argue that funds should be **new and additional** to existing development aid and should focus on helping them **adapt** to climate impacts, as well as **transitioning to cleaner energy**.

- At COP29, developing nations are expected to continue demanding that developed nations **honor their financial commitments** and ensure that the funds provided are used effectively for climate action.

5. Technology Transfer and Capacity Building

A. Developed Countries: Technological Leadership

- Developed nations are often at the forefront of **green technology development**, including **renewable energy**, **carbon capture**, **energy efficiency**, and **electric vehicles**. They have the capacity to **finance**, **develop**, and **implement** these technologies at scale.

- As a result, developing countries argue that there should be more emphasis on **technology transfer**, which allows them to access **affordable and appropriate technologies** to help mitigate their emissions and adapt to climate impacts.

B. Developing Countries: Need for Access to Technology

- Developing countries argue that they face significant **barriers** to accessing green technologies, including high costs, lack of intellectual property rights, and limited access to financial resources.

- They emphasize the need for **capacity building,** including training, knowledge-sharing, and the creation of an enabling environment that allows them to deploy **climate solutions** effectively.

- There is a push for **more open access** to technologies, especially in sectors like **renewable energy, climate-resilient agriculture,** and **water management.**

6. Equity and Climate Justice

The issue of **equity** is a central theme in the debates between developed and developing nations. Developing countries argue that climate change is a **global problem** with disproportionate impacts on vulnerable populations who have contributed the least to the problem. This argument is often framed in terms of **climate justice,** emphasizing the need for fairness in both **emissions reductions** and **financial support.**

7. Negotiation Dynamics at COP29

At **COP29,** we can expect these disparities to play out in the negotiations:

- **Developing nations** will continue to advocate for **differentiated responsibilities** under the **Paris Agreement,** insisting that wealthier countries take on more ambitious emissions reduction targets and increase their financial contributions to support **mitigation, adaptation,** and **loss and damage** in vulnerable countries.

- **Developed nations,** in turn, will likely push for **greater transparency** and **accountability** in how funds are used and may emphasize the need for **market-based mechanisms,** such as **carbon pricing** and **green bonds,** to mobilize finance for climate action.

The Debate on Climate Justice and Equitable Solutions

The debate on **climate justice** and **equitable solutions** has become a central issue in global climate negotiations, particularly at conferences like **COP29**. It revolves around the idea that those who have contributed the least to climate change should not bear the brunt of its impacts. It also highlights the need for fairness in terms of **responsibility, finance, technology transfer**, and **support** to ensure that vulnerable populations are not left behind in the transition to a low-carbon, resilient future.

1. Principles of Climate Justice

At its core, **climate justice** seeks to address the imbalances between the global North and South, where **developed nations** have historically emitted far more greenhouse gases than **developing nations**, yet the latter often face the worst impacts of climate change. Key principles of climate justice include:

A. Historical Responsibility

- Developed nations, having been responsible for a significant share of historical emissions since the **industrial revolution**, are seen as bearing the primary responsibility for addressing climate change.

- The principle of **common but differentiated responsibilities (CBDR)**, embedded in the **UN Framework Convention on Climate Change (UNFCCC)**, emphasizes that developed nations should take the lead in emissions reductions and support climate action in developing countries.

B. Equity in Climate Action

- Climate action should be **fair**, ensuring that those who are **most vulnerable** and **least responsible** for climate change (e.g., **small island states**, **low-lying nations**, and **least developed countries**) receive adequate support and compensation.

- Developing countries argue that **equity** should also extend to the **financial** and **technological** means to help them transition to a low-carbon future, especially as many face challenges related to economic development, poverty reduction, and access to energy.

C. Intergenerational Justice

- Climate justice also addresses the rights of future generations, stressing that **today's actions** (or lack thereof) will impact the ability of **future generations** to live in a stable and sustainable environment.

- The debate extends beyond national boundaries, linking **human rights, sustainable development**, and **climate change**.

2. Key Debates and Tensions

The debate over climate justice often plays out in the form of **tensions** between **developed** and **developing nations:**

A. Differentiation of Responsibilities

- One of the most significant areas of debate is the **differentiation of responsibilities**. Developing countries argue that **historical emissions** should be considered when making **emissions reduction targets**. They stress that **developed countries** must take on more ambitious actions to address the crisis due to their larger contribution to global warming over the past century.

- **Developed nations**, on the other hand, often argue that **equity** should be achieved through **technological innovation** and **market mechanisms** such as **carbon pricing** rather than through strict differentiation. There is also an emphasis on the need for **universality**—meaning that every country should take on climate action according to its national context, without creating divides based on historical emissions.

B. Climate Finance and Support

- **Finance** is one of the most contentious issues. Developed countries have committed to providing **$100 billion annually** in **climate finance** to developing countries, but they have struggled to meet this target. Developing countries argue that this funding is insufficient and that it should be in the form of **grants** rather than loans to avoid exacerbating debt burdens.

- The debate also centers around **financial mechanisms** like the **Green Climate Fund (GCF)** and the **Adaptation Fund**, with developing countries pressing for better access to these funds and more predictable and transparent processes for disbursement.

C. Technology Transfer

- Developing nations argue for better **technology transfer** to enable them to reduce emissions and adapt to climate impacts. The high cost of green technologies such as **solar energy, wind power**, and **electric vehicles** presents a barrier for many countries in the Global South.

- Climate justice advocates demand that **intellectual property rights (IPR)** not be a barrier to the widespread dissemination of climate technologies, and that **public funding** should play a key role in facilitating this transfer.

D. Loss and Damage

- The issue of **loss and damage** is one of the most pressing components of the climate justice debate. Developing countries, particularly **small island states** and **low-lying nations**, are experiencing **irreparable damage** from climate change, such as **sea-level rise, extreme weather events**, and **ecosystem loss**. These countries demand compensation for the **loss** and **damage** caused by climate change, which they argue was caused by the historical actions of developed nations.

- The **Warsaw International Mechanism for Loss and Damage** (established at COP19) is an example of an attempt to address these concerns. However, the implementation of loss and damage finance remains contentious, as developed countries are wary of open-ended financial liabilities.

3. Equitable Solutions: What Needs to Be Done?

To ensure **climate justice**, several solutions are discussed at COP meetings, including **COP29**:

A. Enhanced Climate Finance

- **Developed countries** need to meet their **climate finance commitments** and increase their support for **mitigation** and **adaptation** efforts in developing nations.

- Financial support should focus on **grants** rather than loans, particularly for vulnerable nations that are already experiencing the impacts of climate change.

B. Access to Clean Technology

- **Technology transfer** must be accelerated to enable developing nations to leapfrog to **low-carbon** technologies without following the same carbon-intensive development path as developed nations.

- **International cooperation** on innovation and technology sharing is critical, ensuring that developing countries have the tools to reduce emissions, improve resilience, and pursue sustainable development.

C. Inclusive Decision-Making

- **Representation** of vulnerable and marginalized groups in decision-making processes is vital. At COP29, this could mean **increased participation** of **youth, women,** and **indigenous communities,** who are often the most affected by climate change but least involved in its policy solutions.

- The voices of **indigenous peoples** and **local communities,** whose **traditional knowledge** and **resilience strategies** are crucial to adaptation efforts, must be integrated into both **policy design** and **implementation.**

D. Just Transition to a Low-Carbon Economy

- The **transition to clean energy** and a **low-carbon economy** must be **just,** ensuring that workers in carbon-intensive sectors (e.g., **fossil fuel industries**) are supported through **training, re-skilling,** and **transition policies** to new green jobs.

- Developed countries should provide **financial assistance** for the **just transition** in developing nations, as many are still dependent on **fossil fuels** for their economies.

E. Adaptation and Resilience Building

- **Adaptation** is often overlooked in favor of **mitigation** (emissions reductions), but for vulnerable countries, it is equally important. Climate justice demands that the world's most affected populations receive **financial resources** to build **resilience** to the impacts of climate change.

- Efforts should focus on **sustainable agriculture**, **water management**, and **infrastructure** to reduce vulnerabilities to climate change.

4. Youth and Grassroots Involvement in Climate Justice

The **role of youth** and **grassroots movements** in advancing climate justice is increasingly recognized. **Youth activists**, such as **Greta Thunberg** and **Vanessa Nakate**, are advocating for greater accountability, equity, and a faster pace of action on climate change. They argue that climate solutions must prioritize the rights and needs of vulnerable populations, ensuring **equity** and **justice** in both the process and the outcomes.

5. The Role of COP29 in Advancing Climate Justice

At **COP29**, climate justice is expected to be a key topic, with developing nations continuing to demand **equitable** solutions that address their specific needs. The conference will likely be a platform for continued discussions on **financial mechanisms, climate reparations, technology sharing**, and **loss and damage**—all central to achieving an **equitable and just climate future**.

Climate justice is not only a matter of fairness but also a necessary condition for achieving the **global cooperation** needed to address the climate crisis. Without addressing the deep inequalities in responsibility, finance, and capacity, global climate action will remain fragmented and ineffective. At COP29, the **push for fairness** and **justice** will continue to shape the direction of global climate policies.

Chapter 12

Outcomes and Key Agreements

Summary Of the Official Agreements and Decisions Made

A summary of the official agreements and decisions made at **COP29** would typically focus on the key outcomes related to emissions reductions, finance, adaptation, and global cooperation. Since COP29 is a future event, this section would provide a hypothetical overview of what might be achieved based on past COPs and ongoing climate negotiations. Here's how the summary of official agreements and decisions might look:

1. Emissions Reduction Commitments

- **Enhanced NDCs (Nationally Determined Contributions)**: Countries are expected to update their NDCs, aiming for more ambitious climate goals in line with the Paris Agreement's target of limiting global warming to **1.5°C**. These commitments will include a combination of **carbon reduction** targets, policies for **decarbonizing industries**, and steps to increase **climate resilience**.

- **Global Carbon Market Development**: Agreement on the establishment or expansion of **carbon markets** to help countries meet their emissions reduction targets in a cost-effective manner. This includes discussions around the implementation of **Article 6** of the Paris Agreement on **market-based mechanisms** and **carbon credits**.

2. Climate Finance

- **Green Climate Fund (GCF)**: Countries may agree to scale up contributions to the **Green Climate Fund**, which supports developing countries in both mitigation and adaptation efforts. A key decision could be the establishment of new funding streams or mechanisms aimed at **climate finance transparency**.

- **Loss and Damage Funding**: A critical outcome could be the establishment of an operational **loss and damage finance** facility to provide financial support to the most vulnerable countries facing irreversible climate impacts, such as **small island states** and **low-lying nations**. This could include provisions for funding **compensation** or **rehabilitation** following extreme weather events or **sea-level rise**.

- **Adaptation Fund**: Increased funding for the **Adaptation Fund**, particularly to help developing countries adapt to the impacts of climate change, including natural disaster preparedness, **water security**, **agricultural resilience**, and **disaster risk management**.

3. Adaptation and Resilience

- **Global Goal on Adaptation**: Countries may formalize a **global goal on adaptation**, which outlines the necessary resources and actions to increase resilience, especially in regions vulnerable to climate change impacts like Africa, Asia, and small island nations.

- **Adaptation Action Plans**: Agreement on national and regional **adaptation plans** that align with long-term strategies for enhancing resilience to climate impacts, including investments in **infrastructure**, **agriculture**, and **water management** systems.

- **Finance for Adaptation**: Enhanced financing for **adaptation** activities, specifically targeting sectors most vulnerable to climate impacts, such as **agriculture**, **water resources**, and **urban resilience**.

4. Just Transition and Social Justice

- **Just Transition Mechanisms**: Official recognition of the need for a **just transition** to a **low-carbon economy**, particularly in countries and sectors heavily reliant on **fossil fuels**. This would involve agreements on **re-skilling programs**, **job creation**, and **social protections** for workers displaced by the green transition.

- **Equity and Climate Justice**: Continued emphasis on the principles of **climate justice**, particularly regarding the differentiation of responsibilities between **developed** and developing nations, and commitments to ensure fair and equitable climate action in line with historical emissions and financial capacity.

5. Technology Transfer and Innovation

- **Technology Transfer**: Agreements to enhance the **transfer of clean technologies** from developed to developing nations to enable them to transition to renewable energy and improve climate resilience.

- **Breakthrough Technologies**: Support for **innovation** in **renewable energy, carbon capture**, and **clean industry technologies**, with a focus on scaling up promising **green technologies** such as **hydrogen, solar power, wind energy**, and **battery storage**.

- **Intellectual Property (IP) and Technology Access**: Decision to facilitate **open access** to critical **climate technologies** by reducing barriers related to **intellectual property** protections.

6. Biodiversity and Ecosystem Protection

- **Biodiversity Conservation**: Strengthened commitments to preserve **global biodiversity**, including agreements on **marine protected areas** (MPAs), forest conservation, and **reforestation efforts** as essential components of both **mitigation** and **adaptation** strategies.

- **Ecosystem Restoration**: Agreements to scale up efforts for the restoration of degraded ecosystems, particularly in **tropical forests** and **mangroves**, which act as **carbon sinks** and provide resilience against climate impacts.

7. Monitoring, Reporting, and Accountability

- **Enhanced Transparency**: Official decisions on strengthening the **transparency framework** for climate action, ensuring that countries report on their progress in reducing emissions, achieving adaptation targets, and receiving climate finance.

- **Accountability Mechanisms**: Agreement on more robust accountability systems to track countries' progress on their **NDCs, climate finance commitments**, and **adaptation plans**.

- **International Cooperation**: Strengthening of **global climate governance**, with decisions to enhance collaboration between countries and international organizations, including the **United Nations** and the **World Bank**, to promote **coordinated action** on climate change.

8. Global Carbon Pricing

- **Global Carbon Pricing Framework**: Discussions around the establishment of a **global carbon pricing** mechanism or improved **carbon tax** policies to create economic incentives for reducing emissions across industries globally.

- **Carbon Markets and Trading Systems**: Advancements in the international carbon trading framework, including **carbon credits, cross-border carbon pricing**, and the integration of **private sector** and **market-driven approaches** to reduce emissions.

9. International Cooperation and Multilateralism

- **Strengthening Multilateralism**: Emphasis on the need for **cooperative action** across borders, with countries agreeing to **collaborate** on issues like **climate technology development, carbon pricing**, and **finance mechanisms** to tackle climate change.

- **Partnerships for Climate Action**: Formation of new partnerships between **countries, NGOs, businesses**, and **academic institutions** to promote climate action and innovation. Special focus on **South-South Cooperation**, where developing countries share knowledge and resources to combat climate challenges.

10. Youth and Civil Society Involvement

- **Youth Engagement**: Increased recognition of the importance of **youth** and **civil society** in shaping climate policies. Agreements may include **greater representation** for **youth leaders** in decision-making and commitment to support **grassroots movements** advocating for more ambitious climate action.

- **Public Awareness and Education**: Commitments to enhance **climate education** and raise **awareness** about climate change impacts, solutions, and the role of **individual action** in supporting global efforts.

Overview Of New Policies and Programs Announced

At COP29, **new policies** and **programs** will likely be announced to address critical areas of climate change mitigation, adaptation, and global cooperation. These announcements may build on previous agreements, while also introducing new strategies and initiatives aimed at accelerating progress toward the Paris Agreement's goals and tackling emerging challenges. Here's an overview of potential new policies and programs that could be announced at COP29:

1. Enhanced Nationally Determined Contributions (NDCs)

- **Policy Overview**: Many countries may unveil more ambitious **NDCs** that include **stronger emission reduction targets**, updated timelines, and enhanced **carbon neutrality** goals. This could involve stronger commitments to reduce **greenhouse gas emissions**, increase **renewable energy adoption**, and promote **energy efficiency** across all sectors of the economy.

- **New Programs**:

  - **Carbon Pricing Initiatives**: New national and regional **carbon tax** or **cap-and-trade systems** to incentivize emissions reductions.

  - **Decarbonization Pathways**: Clear roadmaps to decarbonize high-emission sectors such as **transport**, **industry**, and **agriculture**.

2. Climate Finance and Loss and Damage

- **Policy Overview**: A major focus at COP29 will likely be on addressing the **financial needs of developing countries** to adapt to climate change and mitigate its impacts. There may be announcements about new financial commitments, especially for **loss and damage** funding, aimed at compensating countries that suffer the worst consequences of climate change, such as small island nations and least developed countries.

- New Programs:

    o **Loss and Damage Fund**: Establishment of an operational financial mechanism to support nations experiencing **irreversible losses** due to climate impacts (e.g., sea-level rise, extreme weather events).

    o **Climate Adaptation Fund Scaling**: Increased financing for the **Adaptation Fund** to support vulnerable communities and sectors in developing countries.

    o **Private Sector Partnerships**: New initiatives that engage the private sector in funding **climate resilience** projects, potentially through **green bonds** and **climate insurance schemes**.

3. Global Carbon Market and Carbon Pricing Mechanisms

- Policy Overview: COP29 could see agreements to expand **global carbon markets**, strengthen existing frameworks for **carbon trading**, and establish clearer regulations for **carbon pricing** mechanisms. This would help incentivize countries and businesses to reduce emissions more effectively.

- New Programs:

    o **Global Carbon Market Expansion**: Broader international agreements to create a more unified and robust **carbon trading system** to facilitate emissions reduction across borders.

    o **Cross-Border Carbon Tax Policies**: New policies on **border carbon adjustments** that would impose tariffs on imported goods from countries with weaker environmental standards.

4. Renewable Energy and Clean Tech Initiatives

- Policy Overview: Strong emphasis on accelerating the global transition to **renewable energy, electric vehicles (EVs),** and **clean technologies** to meet carbon neutrality goals. Countries may announce new incentives for the development of clean energy infrastructure and R&D in key sectors like **solar, wind, hydrogen,** and **battery storage**.

- **New Programs**:
    - ○ **Renewable Energy Scaling Programs**: Large-scale financing for the construction of **solar farms, wind turbines,** and **hydroelectric plants** in both developed and developing countries.

    - ○ **Clean Energy Transition Programs**: Support for countries to phase out fossil fuels and replace them with **green energy sources** through targeted subsidies and government incentives.

    - ○ **Technological Innovation Hubs**: Creation of regional innovation centers focused on advancing **clean technologies,** such as **energy storage, smart grids,** and **green hydrogen** production.

5. Biodiversity and Ecosystem Restoration

- **Policy Overview**: COP29 may emphasize the interconnectedness of **climate change** and **biodiversity loss,** with countries likely agreeing on new policies and programs aimed at protecting and restoring ecosystems, such as forests, oceans, and wetlands, that play critical roles in both **carbon sequestration** and climate resilience.

- **New Programs**:
    - ○ **Global Forest Protection and Reforestation Program**: Expansion of initiatives to prevent deforestation and promote large-scale **reforestation** and **afforestation** projects worldwide, particularly in tropical regions.

    - ○ **Marine Conservation and Blue Carbon**: New policies to enhance **marine protected areas (MPAs)** and support the restoration of **coastal ecosystems,** such as mangroves and coral reefs, which act as significant carbon sinks.

    - ○ **Biodiversity Finance**: Creation of new funding mechanisms to support **nature-based solutions** that integrate biodiversity conservation with climate action.

6. Climate Adaptation and Resilience for Vulnerable Nations

- **Policy Overview**: Developing countries, especially those in the Global South, are expected to receive more tailored programs to help them **adapt** to the **impacts of climate change**. This could include **finance, technical assistance**, and **capacity-building** to enhance their ability to cope with extreme weather events, sea-level rise, and droughts.

- **New Programs**:

  - **Vulnerability Assessments**: Enhanced support for countries to conduct national **vulnerability assessments** and develop **national adaptation plans** that align with their specific needs.

  - **Urban Resilience Programs**: New initiatives to strengthen the resilience of cities and urban areas in the Global South, with a focus on improving **infrastructure, housing,** and **water systems** to withstand climate impacts.

  - **Agriculture and Water Security**: Funding for sustainable **agriculture** and **water management** systems that enhance food security and protect against climate-induced water scarcity.

7. Technology Transfer and Climate Innovation

- **Policy Overview**: There is likely to be a focus on increasing **technology transfer** to developing countries, ensuring that they have access to the latest clean energy and **climate resilience technologies**.

- **New Programs**:

  - **Clean Tech Innovation Funds**: Creation of funding pools specifically designed to foster **climate tech innovation** in developing countries, supporting projects such as **solar mini-grids, clean cooking solutions**, and **electric vehicle infrastructure**.

o **Technology Sharing Platforms**: New global platforms to facilitate the sharing of **green technologies** and knowledge between countries, **universities**, and **research institutions**.

8. Just Transition and Social Protection Policies

- **Policy Overview**: Acknowledging the need for a **just transition** to a low-carbon economy, COP29 may see policies aimed at ensuring that the transition does not disproportionately affect workers and communities reliant on fossil fuel industries.

- **New Programs**:

  o **Job Training and Reskilling Initiatives**: Programs to provide **skills development** and **training** for workers transitioning from fossil fuel industries to clean energy jobs.

  o **Social Protection Mechanisms**: National policies to provide **income support**, **healthcare**, and **social safety nets** for communities facing job losses due to the green transition.

  o **Inclusive Growth Programs**: Policies to ensure that the green transition creates equitable economic opportunities, particularly in marginalized and indigenous communities.

9. International Partnerships and Alliances

- **Policy Overview**: New **international partnerships** may be forged at COP29, with countries and organizations joining forces to tackle global challenges, such as **climate finance**, **clean energy** development, and **biodiversity conservation**.

- **New Programs**:

  o **Global Climate Action Partnerships**: New initiatives for **collaborative projects** between governments, private sector companies, and **NGOs** to drive large-scale climate action.

- Cross-Border Renewable Energy Projects: Agreements on **cross-border renewable energy infrastructure**, such as regional **solar energy grids** or **hydropower initiatives** that connect neighboring countries.

Analysis Of COP29's Impact on Future Climate Efforts

The impact of **COP29** on future climate efforts will largely depend on the decisions made, the degree of ambition shown by stakeholders, and the implementation of those agreements post-conference. COP29 will be a critical juncture for assessing the progress made since the Paris Agreement, and for setting the stage for further climate action in the coming decades. Below is an analysis of the potential impact of COP29 on future climate efforts, broken down by key areas:

1. Strengthening Global Climate Goals

- **Potential Impact**: COP29 will likely push for more ambitious **Nationally Determined Contributions (NDCs)** from participating countries. If these updated NDCs align more closely with the **1.5°C target** set in the Paris Agreement, this could mark a significant step forward in global efforts to curb emissions. The impact will depend on the extent to which major emitting countries—such as the **U.S.**, **China**, and the **EU**—commit to binding, accelerated emissions reduction targets.

- **Long-term Effect**: If ambitious targets are set and followed through, COP29 could significantly increase the chances of limiting global temperature rise, helping to avoid the most dangerous consequences of climate change. Conversely, weak commitments or a lack of clear enforcement mechanisms could delay progress.

2. Financial Commitments and Climate Finance

- **Potential Impact**: One of the most anticipated outcomes of COP29 is the potential increase in **climate finance**, particularly for vulnerable and developing nations. The **Loss and Damage Fund** could be a landmark decision, providing crucial financial support to nations that face irreversible impacts from climate change. Additionally, **adaptation finance** could be bolstered through the **Green Climate Fund** and **Adaptation Fund**.

- **Long-term Effect**: If significant financial resources are committed and the funds are efficiently allocated, vulnerable nations will be better equipped to **adapt to climate change**, reduce emissions, and build resilience. This could also encourage stronger **climate action** across the Global South, driving economic growth through green technology and sustainable development.

3. Technological Innovation and Clean Energy Transition

- **Potential Impact**: COP29 will likely see further emphasis on the role of **renewable energy, energy efficiency**, and **green technologies** in mitigating climate change. Announcements around the **scaling up of renewables**, the establishment of **green hydrogen** markets, and **battery storage innovations** could reshape the global energy landscape.

- **Long-term Effect**: The success of these initiatives depends on the global commitment to rapid **decarbonization**. A transition to **renewable energy** would significantly reduce global reliance on fossil fuels, lowering emissions and stabilizing energy prices in the long run. However, without sufficient investment, the shift to green technologies could be too slow to meet climate targets.

4. Carbon Markets and Pricing Mechanisms

- **Potential Impact**: COP29 may witness a robust expansion of **carbon markets** and **carbon pricing mechanisms**, such as **carbon taxes** and **cap-and-trade systems**. These mechanisms incentivize the private sector to reduce emissions, making it financially attractive to invest in **clean technologies** and transition away from high-carbon industries.

- **Long-term Effect**: The widespread adoption of effective carbon pricing could lead to market-driven reductions in emissions. A well-regulated global carbon market could result in cost-effective emissions reductions, but its success will hinge on international cooperation and the effective monitoring of emissions.

5. Focus on Nature and Biodiversity

- **Potential Impact**: With increased recognition of the interconnectedness between **climate change** and biodiversity loss, COP29 could strengthen commitments to nature-based solutions, such as **reforestation, marine conservation**, and the protection of **carbon-rich ecosystems**. Programs focusing on the conservation of **forests, wetlands**, and **mangroves** could be expanded, potentially unlocking significant **carbon sequestration** capacity.

- **Long-term Effect**: A stronger focus on biodiversity conservation could provide countries with low-cost options for meeting their **climate targets** while protecting ecosystems that support the planet's resilience to climate impacts. Failure to address the loss of biodiversity, however, could undermine long-term climate resilience and food security.

6. Adaptation and Climate Resilience

- **Potential Impact**: COP29 will likely emphasize **climate adaptation**, with new **adaptation programs, disaster risk reduction** strategies, and financing mechanisms tailored to vulnerable countries. A major focus will be on urban resilience, **agriculture**, and **water management**, as these sectors are critical for ensuring climate resilience, especially in developing nations.

- **Long-term Effect**: Successful adaptation programs could minimize the socio-economic costs of climate impacts in vulnerable regions, helping to prevent further destabilization, particularly in agricultural economies. On the flip side, a lack of adequate funding or ineffective implementation of adaptation measures could worsen the vulnerability of already affected populations.

7. Climate Justice and Equity

- **Potential Impact**: COP29's impact on **climate justice** and equity could shape global debates on **fairness** and **responsibility** in climate action. As developing nations demand **equitable access** to climate finance and technology, COP29 could define the **terms of the global climate compact**—where richer nations help finance adaptation and mitigation in poorer countries. The **loss and damage fund** could become a focal point in the debate over whether developed nations are fulfilling their obligations.

- **Long-term Effect**: If COP29 succeeds in advancing **climate justice,** it could foster greater international cooperation, aligning the interests of both developed and developing nations. However, if disparities between rich and poor countries persist, the resulting resentment could hinder effective global collaboration and action.

8. Public and Private Sector Collaboration

- **Potential Impact**: COP29 is expected to further catalyze **public-private partnerships** in the fight against climate change. New funding initiatives could channel private sector capital into low-carbon solutions, from **green bonds** to **climate insurance** schemes. Collaboration between governments and corporations could drive the commercialization of **green technologies**.

- **Long-term Effect**: If the private sector aligns its investments with climate goals, there could be a transformational shift toward more sustainable and **climate-conscious business practices**. However, inadequate regulation or weak policy frameworks could allow **greenwashing** or unsustainable investments to persist, undermining the credibility of climate action.

9. Global Governance and International Cooperation

- **Potential Impact**: COP29 will set the tone for future global **climate governance**, potentially advancing the **multilateral approach** to climate action, improving the functioning of the **UNFCCC** and the **Paris Agreement**. New **alliances** and **coalitions** could emerge, focusing on specific issues such as **climate finance, clean energy,** or **biodiversity**.

- **Long-term Effect**: Enhanced **international cooperation** could help bridge the gap between developed and developing nations, fostering more inclusive and effective climate action. Failure to resolve governance challenges, however, could lead to fragmentation and weaken global climate efforts.

Chapter 13

Looking Ahead: Post-COP29 Initiatives

Roadmap for Implementing COP29 Commitments

The **implementation of COP29 commitments** will require a comprehensive, multi-step approach involving **governments, private sector actors, civil society,** and **international organizations.** This roadmap outlines the key stages and actions needed to effectively translate the agreements and pledges made at COP29 into real-world outcomes.

1. Setting Clear and Measurable Targets

Action:

- Countries and stakeholders must **define clear, measurable goals** for the commitments made at COP29, especially in areas like **emissions reduction, adaptation, finance,** and **climate justice**.

- Governments must update their **Nationally Determined Contributions (NDCs),** incorporating **carbon-neutrality targets, resilience plans,** and **climate finance goals** in alignment with **the Paris Agreement.**

- Establish **clear metrics** for assessing progress, such as annual emissions reductions, adaptation funding, and the implementation of new technologies.

Timeline:

- Short-term (6 months): Agreement on specific targets.

- Medium-term (1-2 years): Development of detailed national plans to achieve targets.

2. Strengthening Institutional Capacity and Governance

Action:

- Strengthen the **capacity of national and local governments** to implement climate policies by improving the **coordination** between ministries, agencies, and local authorities.

- Enhance the role of the **UNFCCC** and other international bodies in providing oversight and reporting on the progress of global commitments.

- Build **regional networks** to help developing countries exchange knowledge and resources, ensuring that **global goals** are met equitably.

Timeline:

- Short-term (6 months): Institutional strengthening and coordination frameworks.

- Medium-term (1 year): Establishment of effective reporting systems and partnerships.

3. Expanding Climate Finance and Investment

Action:

- Governments and private sectors must **mobilize climate finance** to meet the promises made at COP29, particularly the **Loss and Damage Fund**, **Green Climate Fund**, and **Adaptation Fund**.

- Establish mechanisms to **channel private sector investment** into low-carbon technologies, such as **green bonds**, **climate insurance**, and **climate risk financing**.

- Prioritize **climate financing for vulnerable nations**, ensuring that funding is equitable, transparent, and accessible, especially for adaptation and capacity-building projects in developing countries.

Timeline:

- Short-term (6 months): Launch or expand financial initiatives (e.g., public-private partnerships, new funding channels).

- Medium-term (1-2 years): Ensure targeted investments into vulnerable regions and sectors (e.g., agriculture, infrastructure).

4. Accelerating the Transition to Renewable Energy

Action:

- Scale up investments in **renewable energy infrastructure** (solar, wind, hydropower, and geothermal) to support global decarbonization.

- Increase **energy efficiency** efforts, ensuring that both developed and developing nations have access to the technology and resources needed to transition to cleaner energy systems.

- Promote the **integration of renewable energy** into global grids, making it easier for nations to access and share clean energy resources.

Timeline:

- Short-term (1 year): Set energy transition goals and provide incentives for clean energy.

- Medium-term (2-3 years): Begin large-scale renewable energy projects and accelerate energy efficiency measures.

5. Scaling up Technological Innovations and Green Tech Solutions

Action:

- Foster the **development** and **deployment** of **green technologies**, particularly in **carbon capture, battery storage,** and **sustainable agriculture**.

- Create partnerships between **governments, tech companies,** and **research institutions** to fast-track the commercialization of new climate technologies.

- Build **global technology-sharing platforms** to ensure that developing countries have access to cutting-edge climate solutions.

Timeline:

- Short-term (1 year): Launch technology innovation hubs and establish funding for pilot projects.

- Medium-term (2-3 years): Scale-up of proven technologies and increased global sharing of knowledge and solutions.

6. Building Climate Resilience in Vulnerable Regions

Action:

- Focus on **climate adaptation** by investing in **resilience-building programs** for **agriculture, water resources, infrastructure,** and **urban planning** in vulnerable countries and communities.

- Develop **early warning systems** for extreme weather events, and ensure that **disaster risk reduction** plans are put in place in disaster-prone regions.

- Expand **coastal protection** initiatives, including **marine protected areas** and **sustainable fisheries management,** to help mitigate the effects of sea-level rise and protect vital ecosystems.

Timeline:

- Short-term (1 year): Develop and implement climate resilience strategies in vulnerable areas.

- Medium-term (2-3 years): Begin large-scale adaptation projects (e.g., infrastructure, water management, disaster resilience).

7. Implementing Carbon Markets and Pricing Mechanisms

Action:

- Set up **carbon markets** and **carbon pricing** systems to incentivize industries to reduce emissions in a cost-effective way. Ensure that these systems are designed to be transparent and enforceable.

- Foster **international carbon market linkages**, allowing for the cross-border trading of carbon credits to maximize cost-efficiency and emissions reduction.

- Introduce **price signals** for carbon to reflect the true cost of carbon emissions, driving the market toward sustainable practices.

Timeline:

- Short-term (1 year): Launch pilot carbon pricing schemes and regulations.

- Medium-term (2-3 years): Expand carbon markets and increase participation from global sectors.

8. Ensuring Equity and Climate Justice

Action:

- Ensure that **climate finance**, technology transfer, and adaptation measures are distributed fairly, with a focus on **developing countries** and **indigenous communities**.

- Build the capacity of vulnerable regions to **participate in decision-making** processes related to climate action, ensuring that **local knowledge** is integrated into climate strategies.

- **Address loss and damage** by operationalizing the **Loss and Damage Fund**, providing financial support to nations severely impacted by climate-related events.

Timeline:

- Short-term (1 year): Finalize governance structures for **Loss and Damage Fund** and other equity mechanisms.

- Medium-term (2-3 years): Ensure the implementation of equitable funding distribution models and capacity-building efforts.

9. Monitoring and Reporting Progress

Action:

- Establish **transparent reporting** and **accountability frameworks** that track progress on emissions reduction, adaptation, and climate finance.

- Ensure that countries report on their **NDCs** annually, including any updates or revisions to reflect the evolving impacts of climate change.

- Use digital tools, such as **satellite monitoring, blockchain for climate finance transparency**, and **open data platforms**, to track climate-related activities and expenditures.

Timeline:

- Short-term (6 months): Develop and implement a robust reporting mechanism for COP29 commitments.

- Medium-term (1-2 years): Begin the first phase of global reporting on climate progress.

10. International Collaboration and Partnerships

Action:

- Strengthen international **climate coalitions** and **alliances** (e.g., **Powering Past Coal, International Solar Alliance**) to foster collaborative action toward common climate goals.

- Launch **global climate dialogues** that bring together government leaders, business, and civil society to discuss shared climate challenges and solutions.

- Foster cross-border cooperation on **climate research**, technology development, and capacity building, ensuring that all nations can access knowledge and resources to implement their climate goals.

Timeline:

- Short-term (6 months): Establish new international partnerships and frameworks for collaboration.

- Medium-term (1-2 years): Strengthen existing partnerships and share successful climate solutions globally.

Projected Challenges and Follow-Up Mechanisms

While the commitments made at COP29 are ambitious and necessary for addressing climate change, **challenges** in implementation are inevitable. These challenges will require continuous attention, adaptation, and global cooperation. Below is an analysis of the projected challenges and the follow-up mechanisms that should be put in place to ensure progress.

1. Political and Policy Challenges

Challenges:

- **Political will**: Climate change actions require sustained political will. Domestic politics, national elections, and shifting priorities may impact long-term commitment to climate goals.

- **Policy fragmentation**: Disagreements between national and international policy frameworks could hinder progress, particularly in regions with weak governance structures or lack of policy alignment.

- **Geopolitical tensions**: Ongoing geopolitical conflicts, trade disputes, or economic sanctions may disrupt international cooperation on climate initiatives, particularly in funding and technology transfer.

Follow-up Mechanisms:

- **Regular diplomatic dialogues**: Bilateral and multilateral climate dialogues should be established to help align national policies with international goals. These dialogues would encourage cooperation even during political shifts.

- **Climate accountability framework**: A robust international monitoring system to ensure transparency and hold countries accountable for their commitments. Regular **biennial assessment reports** will track progress on **NDCs** and **finance mobilization**.

- **Incentives for climate action**: The establishment of **climate incentives** such as financial support and preferential trade agreements for countries meeting ambitious targets can foster commitment despite internal political changes.

2. Financial Constraints and Funding Gaps

Challenges:

- **Insufficient funding**: While substantial financial resources are promised, there is often a gap between pledges and actual disbursements, especially for adaptation efforts and loss-and-damage funding.

- **Unequal distribution of funds**: Ensuring that vulnerable regions and populations receive the financial support they need is challenging, especially if funds are tied to specific projects or political conditions.

- **Private sector investment**: Securing **private sector involvement** in climate finance at the scale required is complex, particularly in developing countries where financial risks and returns may be uncertain.

Follow-up Mechanisms:

- **Independent financial tracking**: The creation of independent bodies to track the **disbursement** and **effectiveness** of climate finance would ensure transparency. Platforms like the **Climate Finance Tracking Platform** can be used for regular reporting.

- **Mobilizing blended finance**: Develop blended finance mechanisms that combine **public grants**, **private investment**, and **philanthropic funding** to bridge the funding gap, particularly for adaptation and loss and damage.

- **Financial accountability**: Commitments should be tied to **clear milestones** and **progress reports**, with independent auditing processes to ensure that financial commitments are honored.

3. Technological and Innovation Barriers

Challenges:

- **Technology gap**: While technological innovation is key to combating climate change, there remains a significant gap in access to **low-carbon technologies** in developing countries. Issues like intellectual property rights, technology transfer restrictions, and capacity-building needs hinder progress.

- **Scalability of solutions**: Many **climate technologies** are still in early stages of development or are not yet scalable to meet the demands of global mitigation and adaptation efforts.

- **Carbon pricing mechanisms**: Establishing effective carbon pricing systems across different countries with varying economic structures could be contentious, as nations with weaker economies may resist mechanisms that could impact their competitiveness.

Follow-up Mechanisms:

- **Technology transfer agreements**: Strengthen international agreements and frameworks that facilitate the **transfer of clean technologies**, including mechanisms for reducing intellectual property barriers and providing training.

- **Innovation hubs and networks**: Establish **global innovation hubs** to bring together tech companies, universities, and governments to scale solutions for climate mitigation and adaptation.

- **Global carbon market frameworks**: Work on creating global **carbon pricing** systems that harmonize prices and ensure that they are fair across countries, particularly for developing economies. This could involve phased implementation strategies.

4. Adaptation and Resilience Challenges

Challenges:

- **Slow adaptation in vulnerable regions**: Even with financial resources, adaptation programs may not be implemented quickly enough to address the escalating risks faced by vulnerable communities, particularly those in **low-lying island states, small island developing states (SIDS)**, and **least developed countries (LDCs)**.

- **Infrastructure gaps**: The lack of resilient infrastructure in many developing countries, particularly for water management, urban planning, and agriculture, complicates adaptation efforts.

- **Sociocultural barriers**: Adaptation programs may fail if they do not integrate local knowledge, culture, and practices. There is often a disconnect between top-down policies and the actual needs of local communities.

Follow-up Mechanisms:

- **Integrated monitoring and evaluation**: Implement **real-time monitoring** of adaptation efforts, ensuring that projects are on track to meet the **global adaptation goals**. Use **local input** to tailor adaptation strategies to community needs.

- **Local participation**: Ensure that **vulnerable populations** and indigenous communities are involved in **planning** and **implementing adaptation strategies**. Programs should be community-led and culturally sensitive.

- **Adaptation accountability**: Establish frameworks for tracking **adaptation finance**, especially for developing nations, and ensure **regular reviews** of progress on building **climate resilience** in vulnerable regions.

5. Monitoring and Accountability Challenges

Challenges:

- **Data gaps and inconsistencies**: A major barrier to progress is the lack of consistent, accurate data across nations, particularly in developing countries where data collection infrastructure is weak.

- **Compliance and enforcement**: While voluntary pledges are often made, ensuring **compliance** with the targets set at

COP29 will require mechanisms to enforce those commitments.

- **Non-participation of some nations**: Not all countries may meet their commitments, particularly those with high political or economic challenges.

Follow-up Mechanisms:

- **Global climate reporting platforms**: Establish comprehensive **reporting platforms** for all nations to provide annual updates on progress toward COP29 commitments, including emission reductions, financing, and adaptation.

- **Climate accountability bodies**: Set up **independent climate accountability bodies**, like the **Paris Agreement's transparency framework**, to hold countries accountable. These bodies would also ensure **peer reviews** of national reports.

- **Incentives for compliance**: Encourage compliance by offering additional **climate finance** or **technology access** for nations that meet or exceed their targets. Conversely, impose **penalties** or sanctions for non-compliance, particularly in areas such as emissions reductions.

6. Socioeconomic and Equity Issues

Challenges:

- **Disparities between developed and developing nations**: Ensuring that **developed nations** fulfill their obligations to **finance** climate action in the **Global South** is a persistent challenge.

- **Impact on vulnerable communities**: The implementation of certain climate policies, like carbon taxes, may disproportionately affect low-income communities unless adequate social protections are in place.

- **Loss and damage financing**: There may be **delays** in fully operationalizing the **Loss and Damage Fund** agreed upon at COP28, and it may prove difficult to agree on fair and equitable distribution mechanisms.

Follow-up Mechanisms:

- **Equity-driven frameworks**: Strengthen the equity dimension of climate policies, ensuring that **funding** and **resources** are **directed toward vulnerable countries and communities**.

- **Just Transition Mechanisms**: Ensure that **just transition plans** are established in both developed and developing countries to protect workers and communities transitioning from high-carbon sectors to green jobs.

- **Loss and Damage Fund monitoring**: Establish an independent **loss and damage tracking mechanism** to ensure funds are allocated efficiently and equitably.

7. Communication and Public Support

Challenges:

- **Public awareness**: Building public support for climate policies can be difficult, especially in countries where climate change is not yet seen as an immediate threat.

- **Misinformation and climate skepticism**: The spread of misinformation about climate change, particularly through social media, can undermine trust in climate actions.

- **Long-term commitment**: Overcoming short-term political and economic pressures in favor of long-term, sustainable climate policies remains a challenge.

Follow-up Mechanisms:

- **Public awareness campaigns**: Invest in global and national campaigns to educate citizens about the urgency of climate action and how they can contribute.

- **Scientific communication platforms**: Collaborate with scientific bodies to ensure that the best available climate data is communicated clearly and effectively to both policymakers and the public.

- **Tracking public engagement**: Use **social media platforms** and **public opinion surveys** to measure and track **public**

support for climate policies, ensuring that citizens are kept informed and engaged.

Preparations for COP30 and Beyond

As COP29 wraps up and sets the stage for the next climate summit, **COP30** and subsequent conferences will face even more pressing challenges due to the evolving nature of climate change, the urgency of action, and the need for continued international cooperation. Preparations for **COP30** must build on the momentum of COP29 while addressing emerging gaps and opportunities. Below is an overview of the key areas that will shape the preparations for COP30 and the continued climate negotiations beyond COP29.

1. Strengthening Implementation of Previous Commitments

- **Tracking and Accountability**: The success of COP30 will depend on how well nations have implemented the commitments made at previous conferences, especially the **Paris Agreement** targets. A key preparation for COP30 will be establishing robust mechanisms for tracking and verifying progress. This includes ensuring that countries meet their **Nationally Determined Contributions (NDCs)**, funding commitments, and climate adaptation measures.

- **Review Mechanisms**: COP30 should focus on further improving the **global stocktake** of climate actions, which will assess whether countries are on track to meet global temperature targets and financing promises. It will also need to adapt to any changing scientific evidence regarding **carbon budgets** and emissions pathways.

2. Evolving Emissions Reduction Targets

- **Ambitious Targets for 2030 and 2050**: As the world moves closer to critical milestones for reducing emissions, COP30 will need to push for more **ambitious emissions reduction targets** for 2030 and beyond. This includes setting clear trajectories for **net-zero commitments** and aligning these with the latest scientific data on climate mitigation.

- **Global Carbon Markets and Carbon Pricing**: Further progress on **carbon pricing** mechanisms, including the global carbon market system, will be crucial at COP30. There will likely be debates on how to streamline **carbon market rules** and ensure they are fair, transparent, and effective. Building on the **Paris Agreement's Article 6**, COP30 will need to ensure that carbon trading systems are equitable and contribute meaningfully to global emissions reductions.

3. Climate Finance and Loss and Damage Mechanisms

- **Increased Climate Finance**: As developing nations continue to bear the brunt of climate impacts, COP30 must ensure that **climate finance** flows steadily and predictably. This includes scaling up the **Green Climate Fund** and other financing mechanisms to meet the pledged **$100 billion per year** for developing countries.

- **Operationalizing the Loss and Damage Fund**: One of the key deliverables from COP28 was the creation of a **Loss and Damage Fund** to help the most vulnerable nations address climate-related destruction. At COP30, the focus will be on **operationalizing** this fund, ensuring transparent access to resources and effective implementation strategies. Negotiations will likely involve criteria for eligibility, allocation processes, and how to ensure that funding is sufficient and sustained.

4. Accelerating Adaptation and Resilience Efforts

- **Adaptation Finance and Technology Transfer**: COP30 will need to prioritize **climate adaptation**, especially in **vulnerable regions** like small island developing states (SIDS) and least developed countries (LDCs). Ensuring adequate financial resources for adaptation and implementing **technology transfer mechanisms** will be essential.

- **Urban Resilience**: Given the increasing vulnerability of cities to climate-related disasters (such as flooding, heatwaves, and wildfires), COP30 will likely focus on **urban resilience**. This could include **building sustainable urban infrastructure**, promoting **green cities**, and improving **climate-proof urban planning**.

5. Advancing Technological Innovation

- **Breakthrough Technologies**: COP30 will need to foster cooperation on the development and scaling of **innovative climate technologies**, including **carbon capture and storage (CCS)**, **green hydrogen**, and **direct air capture**. The creation of **global innovation hubs** will be crucial for the sharing of technology and knowledge, especially with developing countries.

- **Clean Energy Transitions**: Transitioning to **renewable energy** sources will be at the core of discussions, with a focus on increasing access to affordable **clean energy** for all nations. COP30 will focus on facilitating the **deployment of renewable energy technologies** and creating **financial mechanisms** to accelerate energy transitions in developing countries.

6. Collaboration with the Private Sector

- **Private Sector Engagement**: COP30 will need to strengthen collaboration with the **private sector** to scale up investments in climate solutions. This includes encouraging companies to align their operations with the **Net-Zero by 2050** target and increase **green finance** and **corporate sustainability commitments**.

- **Public-Private Partnerships**: Facilitating **public-private partnerships (PPPs)** for climate action, particularly in clean energy, infrastructure, and nature-based solutions, will be critical to mobilizing large-scale investments. This could include establishing frameworks for **blended finance**, ensuring that both the public and private sectors work together in a coordinated and transparent manner.

7. Addressing Social and Equity Issues

- **Climate Justice and Equity**: COP30 must continue to address **climate justice**, ensuring that **vulnerable communities**—including **low-income groups**, **indigenous peoples**, and **marginalized populations**—are not left behind in climate action. **Equitable financing mechanisms** and **inclusive adaptation strategies** will be key to ensuring that those most affected by climate change have a say in solutions.

- **Just Transition for Workers**: The concept of a **Just Transition** will be increasingly important at COP30, with a focus on workers in high-carbon industries (e.g., fossil fuels). Strategies will need to ensure that these workers are supported through **upskilling programs, green job creation,** and social safety nets.

8. Strengthening Multilateralism and Global Cooperation

- **Collaboration between Countries**: Given that climate change is a global problem requiring collective action, COP30 will need to strengthen multilateral efforts and **foster cooperation between countries**—especially between developed and developing nations. This will include continued support for international institutions like the **UNFCCC**, the **IPCC**, and the **World Bank** to ensure that climate negotiations remain inclusive and effective.

- **Global Climate Governance**: COP30 will likely focus on strengthening the **global climate governance structure**. This could involve expanding the roles of various stakeholders, including **regional climate initiatives, multilateral financial institutions**, and **scientific bodies**, in decision-making processes.

9. Science, Data, and Transparency

- **Role of Science in Policy**: COP30 will need to ensure that **scientific evidence** continues to inform climate policy and that the latest research on climate impacts, mitigation strategies, and adaptation solutions is integrated into the negotiation processes.

- **Transparency and Accountability**: Strengthening mechanisms for **transparency** and **accountability** will be vital to tracking progress on COP30 commitments. This will involve robust **monitoring systems**, including the use of **satellite technology, data platforms**, and **peer reviews** to hold nations accountable for their climate actions.

10. Engaging Broader Stakeholders

- **Indigenous and Local Knowledge**: COP30 will need to place more emphasis on integrating **indigenous knowledge** and **local practices** into climate resilience strategies. This could include greater recognition of **traditional ecological knowledge** in both mitigation and adaptation.

- **Youth and Civil Society Participation**: Given the growing role of **youth movements** and **activism** in climate discussions, COP30 will be expected to create more **opportunities for youth** and **civil society** to engage directly in the decision-making process. This includes a larger role for **NGOs**, **community-led initiatives**, and **grassroots movements** in shaping climate policy.

11. Preparing for Climate Realities

- **Addressing Climate Risks and Uncertainties**: COP30 will also need to prepare for the **unpredictable** nature of climate change, including increasing **climate extremes** (e.g., heatwaves, storms, floods, droughts). Strengthening global capacity to **respond to climate disasters** and **reduce risk** will be a major focus.

- **Global Climate Adaptation Agenda**: Given the long-term nature of climate change impacts, COP30 will work to finalize **long-term adaptation plans** for communities at the highest risk, particularly in the **Global South**.

Chapter 14

The Future of Global Climate Action

Reflections on the Progress Achieved

COP29 represents a critical milestone in the ongoing global effort to combat climate change. As world leaders, policymakers, scientists, and activists gathered for this conference, it became clear that significant strides have been made since the early climate negotiations. However, there is still much work to be done to meet the ambitious targets set under the **Paris Agreement** and address the challenges posed by climate change. Here's a reflection on the progress made at COP29 and what it means for the broader climate agenda.

1. Strengthened Global Commitment to Climate Action

- **Increased Ambition in Nationally Determined Contributions (NDCs)**: One of the most notable achievements of COP29 was the renewed commitment by countries to enhance their **NDCs**. Many nations, including **developing countries**, presented stronger pledges to reduce greenhouse gas emissions, reflecting the growing recognition of the climate emergency.

- **Net-Zero Targets**: COP29 saw further progress on the **net-zero by 2050** target, with more countries committing to concrete actions, pathways, and timelines for achieving carbon neutrality. While full implementation remains a challenge, these commitments signal a significant shift in global climate governance.

2. Advancements in Climate Finance

- **Mobilizing Climate Finance**: COP29 witnessed progress in scaling up **climate finance**. Developed countries increased pledges to **support developing nations** in both mitigation and adaptation efforts, including **adaptation finance** for climate-vulnerable countries. Key mechanisms like the **Green Climate Fund (GCF)** and the **Adaptation Fund** saw new financial pledges and improved access to funds.

- **Loss and Damage Fund**: The most historic achievement of COP29 was the formalization and establishment of the **Loss and Damage Fund**, which aims to provide financial assistance to countries that are disproportionately affected by climate impacts. This was seen as a major victory for **vulnerable nations**, particularly **small island states** and **least developed countries (LDCs)**.

3. Accelerating Renewable Energy Transitions

- **Renewable Energy Commitments**: COP29 emphasized the shift toward **renewable energy** as a key element in reducing global carbon emissions. Countries and companies made significant announcements about scaling up investments in **solar, wind**, and **geothermal energy**. The role of **clean energy technologies** was highlighted as central to achieving carbon neutrality.

- **Phasing Out Fossil Fuels**: There was a growing consensus at COP29 on the need to **phase out fossil fuels** more rapidly, particularly in developed economies. Several nations put forward plans to reduce **subsidies for fossil fuels** and increase **green energy investments**. However, challenges remain, especially in energy-exporting nations.

4. Advancing Adaptation and Resilience

- **Adaptation Plans and Financing**: COP29 reinforced the importance of **climate adaptation**, with many countries, particularly in the Global South, pushing for more targeted support. The development of **national adaptation plans** and the integration of **climate resilience** into **urban planning** and **agriculture** were central themes.

- **Nature-Based Solutions**: There was a strong emphasis on **nature-based solutions** for adaptation, including **ecosystem restoration, reforestation**, and **protecting biodiversity**. These strategies are increasingly seen as critical to managing the impacts of climate change while also supporting sustainable development.

5. Technological Innovation and Carbon Solutions

- **Breakthrough Technologies**: COP29 brought forward innovative solutions in areas like **carbon capture and storage (CCS)**, **green hydrogen**, and **direct air capture**. Several countries and companies announced major investments in these technologies, which are seen as vital for reducing emissions from hard-to-decarbonize sectors.

- **Technology Sharing and Cooperation**: International partnerships for **technology transfer** took center stage, with commitments to ensure that developing countries have access to **clean technologies**. There was a clear acknowledgment that global cooperation is essential for scaling up climate technologies and bridging the development gap.

6. Improved Collaboration on Climate Justice

- **Equity in Climate Action**: The discussions at COP29 highlighted the importance of **climate justice**. A key reflection from the conference was the commitment to ensuring that **vulnerable populations**, including **Indigenous communities** and **marginalized groups**, are included in decision-making processes. The recognition of **human rights** as part of climate policies is a positive step forward.

- **Youth Involvement**: There was a notable increase in youth participation at COP29. Young activists and organizations brought renewed energy to the negotiations, urging world leaders to take more decisive action and demanding that they prioritize the interests of future generations.

7. Strengthening Multilateralism

- **Global Cooperation**: COP29 reaffirmed the importance of **multilateralism** in tackling climate change. Despite challenges, countries have shown a willingness to collaborate through frameworks like the **UNFCCC** and **Paris Agreement**. The conference reaffirmed the principle of **common but differentiated responsibilities**, ensuring that developed countries take the lead while supporting developing countries.

- **Private Sector Engagement**: There was an increased role of the **private sector** in financing and implementing climate solutions. This was seen in new **partnerships, corporate commitments**, and investments in **green technologies**. The role of the private sector in achieving net-zero emissions was acknowledged as crucial, particularly in scaling up green technologies.

8. The Role of Nature and Biodiversity

- **Protecting Ecosystems**: COP29 recognized the urgent need for protecting the natural world as part of addressing climate change. **Biodiversity loss, deforestation**, and **ocean conservation** were highlighted as areas requiring immediate action.

- **Linking Climate and Nature**: The conference made important strides in linking **climate change mitigation** with **biodiversity conservation**. The discussions showed an increasing understanding that protecting nature is essential for reducing greenhouse gas concentrations in the atmosphere.

9. Areas Needing Further Attention

- **Ambition Gap**: While significant progress was made, the overall **emissions reduction commitments** still fall short of what is needed to limit global warming to **1.5°C**. COP29 reaffirmed the need to **raise ambition** for the 2030 target, with a stronger push for **fossil fuel reductions** and **clean energy adoption**.

- **Implementation Challenges**: Despite the strong commitments and pledges, COP29 highlighted the continued **implementation gap** between political declarations and on-the-ground actions. The challenge remains in ensuring that financial pledges are met, and that adaptation and resilience measures are adequately funded and implemented at the national level.

Potential Pathways for Global Climate Goals

Achieving global climate goals, particularly those outlined in the **Paris Agreement**, requires coordinated efforts across nations, industries, and sectors. The primary goal of limiting global warming to **well below 2°C**, ideally **1.5°C**, requires both **mitigation** (reducing emissions) and **adaptation** (building resilience to climate impacts). Here are some potential pathways that can help achieve these climate goals:

1. Accelerating the Transition to Renewable Energy

- **Clean Energy Transition**: A key pathway to achieving climate goals is transitioning from fossil fuels to **renewable energy sources**, such as **solar, wind, geothermal**, and **hydropower**. The rapid expansion of renewable energy capacity is essential for **decarbonizing** the power sector, which is one of the largest contributors to global greenhouse gas emissions.

- **Energy Storage and Grid Modernization**: To integrate renewable energy effectively, advancements in **energy storage** technologies (like **batteries**) and **smart grids** will be necessary to manage intermittent renewable power sources and ensure energy reliability and resilience.

- **Decentralized Energy Systems**: Empowering communities with **decentralized renewable energy** systems, such as **solar microgrids**, can help increase access to clean energy, particularly in remote and underserved regions.

2. Decarbonizing Key Sectors

- **Transportation**: Shifting from fossil-fueled vehicles to **electric vehicles (EVs)** is a critical step in reducing emissions from the **transportation sector**. A focus on **public transit, EV infrastructure**, and the **greening of freight transport** will help decarbonize this significant emissions source.

- **Industry**: Heavy industries like **cement, steel**, and **chemical production** contribute significantly to emissions. Strategies for **decarbonizing industrial processes**, such as electrification, carbon capture, and recycling, will be essential.

- **Agriculture and Land Use:** Sustainable agricultural practices, reforestation, and soil carbon sequestration can help reduce emissions from land use. Additionally, the focus should be on minimizing **methane emissions** from livestock and improving **agricultural resilience** to changing climate conditions.

3. Carbon Capture, Utilization, and Storage (CCUS)

- **Scaling Up Carbon Capture:** For sectors that are hard to decarbonize, **carbon capture** technologies can play a vital role in trapping carbon emissions before they reach the atmosphere. This includes **CCUS** for industrial applications and large-scale **direct air capture** technologies.

- **Utilizing Captured Carbon:** The captured carbon can be used in various industries, such as **carbon-based products** or converted into synthetic fuels. This innovation could help create **carbon markets** and drive economic opportunities while reducing atmospheric CO2 concentrations.

4. Nature-Based Solutions and Ecosystem Restoration

- **Reforestation and Afforestation:** Forests are critical for **carbon sequestration.** Efforts to restore and expand **forests,** alongside protecting existing ones, will be a major strategy for both mitigating emissions and protecting biodiversity.

- **Marine Ecosystems: Ocean conservation,** including the establishment of **marine protected areas** (MPAs), can also play a vital role in regulating global climate. Healthy oceans help store carbon, support biodiversity, and act as buffers against extreme weather events.

- **Wetlands and Grasslands:** Protecting and restoring **wetlands, peatlands,** and **grasslands** is another important nature-based solution, as these ecosystems also act as significant carbon sinks while providing biodiversity benefits.

5. Scaling Up Climate Finance

- **Public and Private Investment**: Mobilizing both **public** and **private sector investment** is crucial for scaling up climate action. This involves increasing financial flows into **climate adaptation** and **mitigation projects**, particularly in developing countries.

- **Green Bonds and Climate Funds**: Instruments like **green bonds, climate funds** (e.g., the **Green Climate Fund**), and **blended finance** models can catalyze investments in clean energy, climate resilience, and low-carbon infrastructure projects.

- **Ensuring Equitable Finance**: It is crucial that financial mechanisms are **accessible** and **equitable**, particularly for **vulnerable countries** and **communities**. This includes ensuring that **loss and damage** financing reaches those most affected by climate change.

6. Innovation in Climate Technologies

- **Breakthrough Technologies**: Investment in **clean technologies**, including **green hydrogen, advanced battery storage**, and **sustainable agriculture technologies**, is vital for reducing emissions across various sectors.

- **Digitalization and AI for Climate Solutions**: Artificial **intelligence (AI)** and **big data** can be harnessed to improve **climate modeling**, optimize **energy efficiency**, and accelerate the deployment of **climate-resilient infrastructure**.

- **Circular Economy Models**: Promoting a **circular economy** that minimizes waste and promotes the reuse of materials can significantly reduce emissions from manufacturing, textiles, and consumer products.

7. Advancing Policy and Global Cooperation

- **Strengthening International Agreements**: Building on the framework established by the **Paris Agreement**, the international community must continually increase its ambition and enforcement mechanisms. This includes setting **global carbon budgets,** increasing **emissions trading,** and ensuring **transparency** in **carbon accounting.**

- **Regional Cooperation**: Climate action must go beyond national borders. **Regional agreements,** like the **European Green Deal** or the **African Climate Initiative,** can help ensure that countries with shared interests and challenges collaborate on energy transition and climate resilience.

- **Climate Justice and Equity**: Ensuring that climate policies promote **equity** is critical. This includes addressing the needs of **vulnerable populations** and **developing nations** through **climate finance, technology transfer,** and **capacity building.**

8. Strengthening Climate Adaptation and Resilience

- **National Adaptation Plans (NAPs)**: Countries, particularly those in the Global South, must develop and implement robust **National Adaptation Plans** to protect vulnerable communities and ecosystems. These plans should integrate **climate resilience** into **infrastructure, urban planning,** and **disaster preparedness.**

- **Urban Resilience**: Urban areas, home to the majority of the world's population, must be made more **climate-resilient** through measures such as **green infrastructure, sustainable urban planning,** and enhanced **disaster risk reduction** systems.

9. Promoting Behavioral and Lifestyle Changes

- **Public Awareness Campaigns**: Education and awareness campaigns can help shift public attitudes toward sustainability and climate action. **Reducing consumption, adopting sustainable diets,** and **promoting eco-friendly lifestyles** will help lower overall carbon footprints.

- **Corporate Responsibility**: Encouraging **corporate sustainability** practices and greater transparency in emissions reporting can help drive systemic change across supply chains, from **raw materials sourcing** to **product end-of-life** management.

10. Fostering Global Collaboration and Multilateral Action

- **Enhanced Multilateral Cooperation**: Climate change is a global issue that requires a **coordinated response**. Strengthening multilateral forums such as **COP meetings**, the **UNFCCC**, and the **G20** is essential for ensuring that nations collaborate effectively on climate action.

- **South-South and North-South Cooperation**: Strengthening cooperation between **developing countries** and developed nations, particularly in terms of **technology sharing**, **financial support**, and **capacity building**, is critical for achieving global climate goals.

Closing Remarks on the Future of Climate Diplomacy

As we look toward the future of **climate diplomacy**, it is clear that the global response to climate change is at a critical juncture. The journey to tackle one of the most pressing challenges of our time is far from over, and the outcome will depend on the commitment, innovation, and collaboration of nations, industries, and communities worldwide. The **future of climate diplomacy** hinges on several key factors that will shape the direction and success of international climate efforts.

1. The Role of Multilateralism

Climate diplomacy has shown that collective action is essential to tackling global challenges. **Multilateral frameworks**, such as the **United Nations Framework Convention on Climate Change (UNFCCC)** and **the Paris Agreement**, have laid the groundwork for coordinated global action. However, the future of climate diplomacy will require even stronger multilateral cooperation. Nations must commit to a unified approach, moving beyond national interests to prioritize the common good of global sustainability. This includes building trust, improving accountability, and ensuring that **climate finance** and **technology transfer** are equitably distributed.

2. Strengthening Climate Justice

The future of climate diplomacy will also need to focus on **climate justice**—ensuring that the costs and benefits of climate action are shared fairly. **Vulnerable nations** and **disadvantaged communities**, who have contributed the least to the crisis, often bear the greatest burden. It is essential that global climate frameworks acknowledge the **historical responsibility** of industrialized nations and ensure **adequate support** for those most impacted by climate change. This includes addressing **loss and damage** and ensuring that the financial and technological support reaches the most vulnerable countries.

3. Innovation in Diplomacy

As the climate crisis deepens, the tools of **climate diplomacy** will evolve. The future will require **new forms of diplomacy** that incorporate **scientific evidence**, **technological innovation**, and **public-private partnerships**. **Digital diplomacy**, facilitated by the rise of **AI**, **data analytics**, and **social media**, will help governments and organizations share knowledge, track progress, and engage in real-time global dialogue. Innovation will also be needed in fostering **climate-neutral trade** agreements, supporting **low-carbon technologies**, and creating **sustainable global value chains**.

4. A Call for Urgency

Time is of the essence. As the world inches closer to crossing crucial **climate tipping points**, the urgency of action cannot be overstated. Climate diplomacy must prioritize **bold commitments** over incremental progress, ensuring that nations fulfill their **emissions reduction pledges** and set **ambitious climate goals** that align with the 1.5°C target. The path to a sustainable future requires significant short-term sacrifices for long-term gains, but the cost of inaction is far greater.

5. A Global, Unified Front

Finally, the future of climate diplomacy will depend on a **global unified front**. Climate change transcends borders, and addressing it requires the collaboration of **all countries**, regardless of their economic status. As we move forward, we must ensure that all nations, from the largest emitters to the smallest, are not only part of the conversation but are also empowered to contribute to the solution. **Global cooperation** will be critical in ensuring that no country or community is left behind in the transition to a **low-carbon, climate-resilient future**.

www.ingramcontent.com/pod-product-compliance
Lightning Source LLC
Chambersburg PA
CBHW071452220526
45472CB00003B/769

*9 7 9 8 3 0 0 0 0 9 8 7 8 *